Genetics Is Easy

FOURTH REVISED

AND

ENLARGED EDITION

BY PHILIP GOLDSTEIN

Chairman, Department of Biology
Abraham Lincoln High School
Brooklyn, N. Y.

1967

LANTERN PRESS • PUBLISHERS

NEW YORK, N. Y. 10010

FOURTH REVISED EDITION 1967

Library of Congress Catalog Card Number 67-12011

PUBLISHED SIMULTANEOUSLY IN CANADA BY
GEORGE J. MCLEOD, LIMITED, TORONTO

MANUFACTURED IN THE UNITED STATES OF AMERICA

GENETICS IS EASY

This edition of *GENETICS IS EASY* is dedicated to

GREGOR MENDEL

on the 100th anniversary of the publication of
his classic paper entitled
EXPERIMENTS IN PLANT HYBRIDIZATION

TABLE OF CONTENTS

INTRODUCTION TO THE FOURTH EDITION

IN THE PAST two decades, genetics has entered a fascinating phase in which sophisticated questions on the chemical and physical nature of genes and gene action are often asked, and sometimes even answered. The geneticist is now probing into the molecular events involved in the reproduction and expression of genes. These investigations have brought about a shift in the level of abstraction of genetics from the cellular level to the molecular level. Nowhere is this more apparent than in the current terminology of molecular genetics, where common genetic terms such as *allele* and *linkage* have been complemented by such formidable terms as *muton, recon, cistron, operon, codon,* and *anticodon.*

The resemblance of some of these new terms to the particles of modern physics is not a coincidence, but rather is a reflection of the high flux of physicists in the field of genetics. This transition from physics to biology was not accomplished without misgivings, however, as witnessed by the wistful analogies to nuclear physics. The rumors that some of these terms are a result of confusing the nucleus of a cell with that of an atom are probably unfounded. Terminology aside, the development of genetics is remarkably similar to that of physics, one development leading logically to another, in each case pressing toward even more fundamental particles. In treating a complex subject such as genetics, it is often simpler to omit much that is historical and classical and to concentrate on topics that are currently exciting. Such omissions are a disservice to the reader, since much of the flavor of the subject would be lost. I am pleased therefore, that Mr. Goldstein has retained the classical material in the fourth edition of *Genetics Is Easy,* and has chosen to add chapters on molecular genetics. He has succeeded in presenting the material of an increasingly recondite field with refreshing clarity, eliciting from me, for instance, a frank envy of his discussion of the genetic code.

The aficionado of genetics might criticize the omission of genetic regulatory mechanisms in this edition, in particular the concept of the *operon* advanced by the bacterial geneticists of the Pasteur Institute. I personally agree with the author's decision, not only because the hypotheses involved are still evolving and being refined, but also because most of the experimental work relating to this topic has been confined to microbial systems. Indeed, there is some reason to believe that the *operon* hypothesis, if valid, applies only to bacteria.

In this day of moon probes and neutrinos, the value of an author who can summarize the knowledge in a scientific discipline and communicate this knowledge to the uninitiated reader cannot be measured except, perhaps, in terms of the number of readers that he stimulates sufficiently to seek more information on the subject. I strongly suspect that Mr. Goldstein's worth would not suffer if measured by that criterion.

A. S. L. Hu, Ph.D.
DEPARTMENT OF BIOCHEMISTRY
UNIVERSITY OF KENTUCKY,
LEXINGTON, KY.

GENETICS IS EASY

CHAPTER 1

BEFORE MENDEL

GENETICS WAS NOT always a science. Its origins are shrouded in the mysterious darkness of the past. Its roots go back through countless generations of farmers and herders whose trial-and-error efforts led to the improvement of crops and flocks. It stems from the questions, the traditions, the folklore, and the superstitions of thousands of generations of people all over the world, whose blind groping, accidental observations, and planned experiments inevitably led to the accumulation of knowledge from which the science of genetics emerged.

What is genetics? It is the science which attempts to clarify relationships among living things. It is the science which tries to explain why living things resemble their parents, and yet differ from them. It is the science which differentiates between what is inherited and what is not. In brief, genetics is the study of heredity and variation.

Since heredity and variation result from reproduction, genetics could not possibly attain status as a science until reproduction was clearly understood. Although the ancients knew a great deal about reproduction, their concepts were vague and often incorrect. For example, even five thousand years ago, the Babylonians and Assyrians carried out the process of pollination to make their date palms produce fruit. But because they did not understand what they were doing, they turned to the mysterious and supernatural for an explanation. Sexual reproduction in plants continued to remain a mystery until 1694, when it was finally cleared up on a scientific basis by the work of Rudolph Camerer. Animal reproduction was also a mystery to our ancient ancestors, even though they knew that animals mated before offspring were produced. Sperms were unknown until 1670, and even after they were discovered their function was not understood. As a matter of fact, the theory that sperms were parasites in the body persisted until about 1840 when it was shown that they were products of the male animal. The actual entrance of a sperm into an egg was not observed until 1854.

So it is evident that science was not ready for a real study of heredity until little over one hundred years ago. At about that time many breeding experiments were reported from all parts of the world. Since these experiments were all fundamentally similar, only those of one man, John Goss of England, will be described here. Goss reported some interesting facts about inheritance in peas. His observations are shown in Fig. 1.

Unfortunately, Goss did not analyze his experiment statistically, nor did he leave us any interpretation of the results. Probably this caused him to overlook the fundamental laws of inheritance which fairly cried out to be discovered in his work. Other workers of the same period also failed to grasp the significance of their experiments, and so it remained for Gregor Mendel to enunciate the basic principles of heredity in 1865.

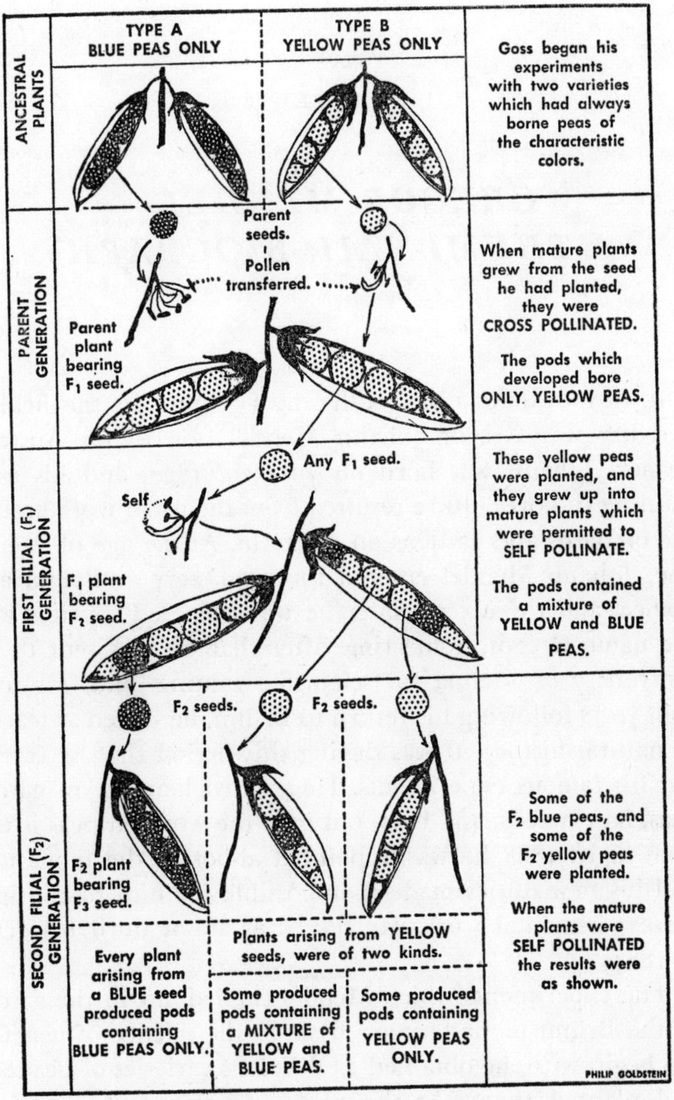

FIG. 1. Diagrammatic representation of the experiments with peas reported by John Goss in 1822. (*Color refers to color of cotyledon.*)

CHAPTER 2

GREGOR MENDEL— A THUMBNAIL BIOGRAPHY

THE OUTSTANDING nineteenth-century figure in the field of heredity was Gregor Johann Mendel, son of an Austrian farmer. Johann was born on July 20, 1822, and his early training in horticulture resulted from the usual work around the orchards and gardens on the farm. At the age of twenty-one, Johann Mendel entered the monastery at Brünn as a novice, and three years later he was ordained a priest with the name Gregor. Some time after that, he was sent to the University of Vienna for scientific training, and, for fourteen years following his return to Brünn, he served as teacher of natural history. It was during this period that he carried out his famous experiments. He studied heredity in garden peas, hawkweeds, and bees, but only the work on peas is well known. In 1868 he was appointed abbot of the monastery, and his new duties made it impossible for him to continue his experimental work. He served as abbot until his death in 1884.

The experiments which Mendel carried out in the garden of the Brünn monastery are basic to the science of genetics. To begin with, he obtained 34 different varieties of pea seeds and planted them. As the plants developed, he carefully studied their characteristics, making sure that every one of the varieties bred true to its type. After comparing the characteristics of the 34 varieties he now had growing, Mendel selected for his experimentation those which showed definite contrast in certain specific traits, eliminating any in which the difference was not clear enough to permit of certain separation. In all, he selected 7 traits (or *unit characters,* as he called them) with which to work. These traits are listed in Fig. 2, together with the alternative ways in which they appeared in different varieties of pea plants (see also Fig. 4).

Mendel's Seven Unit Characters	*Alternate Ways in Which Each Unit Character Showed Itself*
1. Seed shape	Ripe seeds were either smooth and round, or angular and wrinkled
2. Cotyledon color	Seed leaves (cotyledons) were either yellow or green in color
3. Seed-coat color	Seed coats were either white or gray
4. Pod shape	Ripe pods were either constricted between the seeds or not so constricted
5. Pod color	Unripe pods were either green or yellow in color
6. Pod (flower) position	The flowers and pods were distributed *axially,* all along the stem, or they were bunched *terminally* at the top of the stem
7. Stem length	The stem was either tall (6-7 feet) or short (¾-1½ feet)

FIG. 2. The seven unit characters with which Mendel worked in his experiments with pea plants.

Having done this preliminary work, Mendel was now ready to begin his actual experimentation. A summary of these experiments is presented in the next section.

FIG. 3. Gregor Mendel and the allelomorphs which he studied in pea plants.

CHAPTER 3

MENDEL'S FAMOUS EXPERIMENTS

MENDEL'S FAMOUS experiments with peas were reported in a paper entitled *Experiments in Plant Hybridization* in the *Proceedings of the Natural History Society* (Discussion IV, 1866), in Brünn. These experiments will be described here in simplified form.* For convenience, the experiments will be considered in three groups, and it will be assumed that each group of experiments was performed to answer a specific question. Wherever possible, the experiment will also be presented diagrammatically.

It was previously pointed out that the pea plants Mendel used bred true to type, so that a pea plant bearing yellow seeds always produced other yellow-seeded plants, and a green-seeded pea plant invariably gave rise to more green-seeded plants. Now Mendel raised the following question.

* A more complete analysis of Mendel's experiments may be found on pages 127-147 of the book, *TRIUMPHS OF BIOLOGY,* by Philip Goldstein, Doubleday and Co., New York, 1965. Reprints of the complete paper (English translation) may be obtained from the Harvard U. Press, Cambridge, Mass.

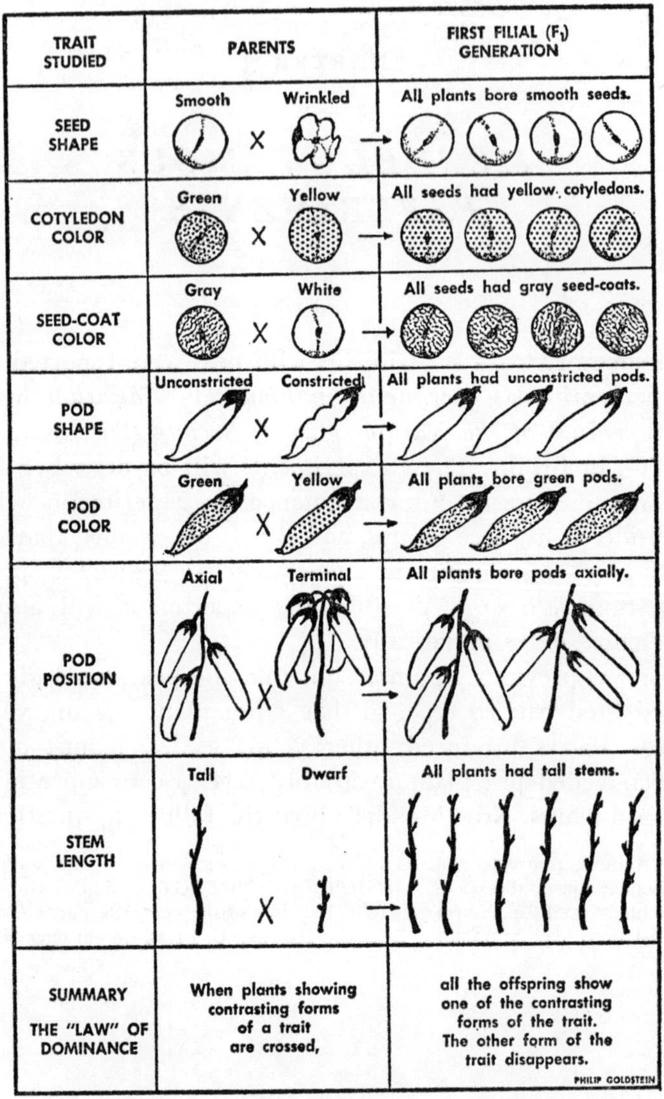

FIG. 4. Results of Mendel's "first set of experiments."

Question 1 ***What sort of offspring will result from the artificial cross-fertilization of plants differing in a single characteristic?***

For example, if a tall variety is fertilized with the pollen from a short variety, what kind of offspring will be produced? Theoretically, several different results are possible. All the offspring might turn out to be intermediate between the two parents; or all the offspring might be tall; or they might all be short; or some might be tall while others were short; etc. What would actually happen if the experiment were done? Mendel performed the crosses indicated in Fig. 4, and his results are shown diagrammatically.

From these results, it appears that one characteristic of each pair exerts more influence in a cross than does its opposite member. Mendel used the word *dominant* to refer to the characteristic which was more potent, and the word *recessive* to refer to the characteristic which disappeared. Thus tall stem is dominant over short stem, and green cotyledon is recessive to yellow. He referred to the plants which resulted from the crossing of the two original pure strains as *hybrids.*

One striking thing about the experiments just described is the total disappearance of the recessive trait in the F_1 generation. This led directly to the second question which Mendel tried to answer.

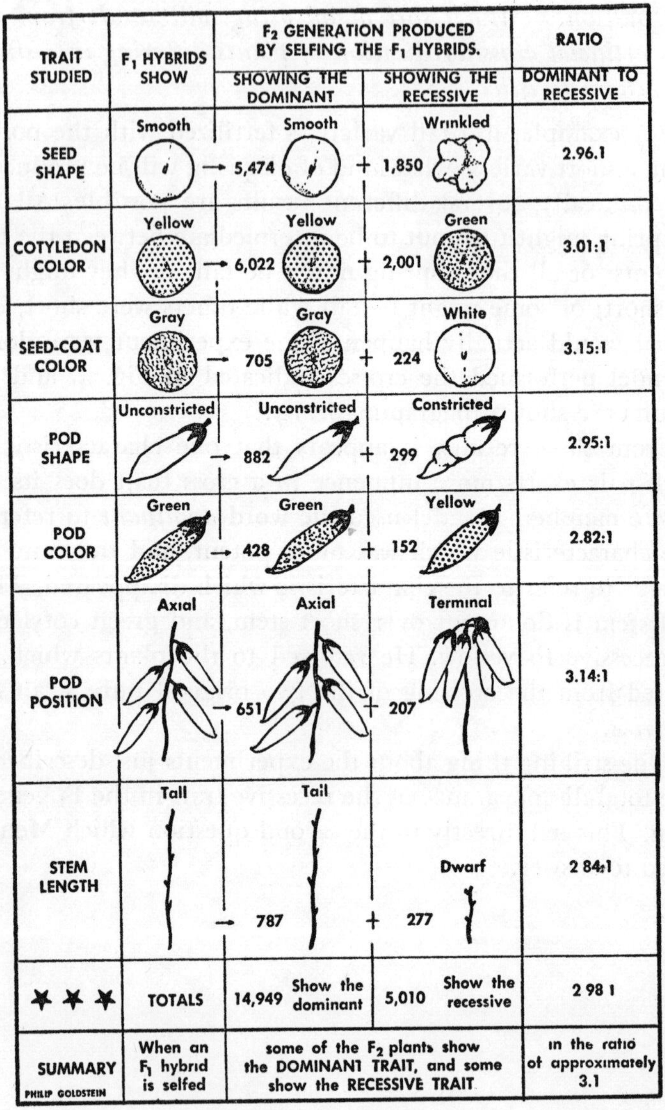

TRAIT STUDIED	F_1 HYBRIDS SHOW	F_2 GENERATION PRODUCED BY SELFING THE F_1 HYBRIDS. SHOWING THE DOMINANT	SHOWING THE RECESSIVE	RATIO DOMINANT TO RECESSIVE
SEED SHAPE	Smooth	Smooth 5,474	Wrinkled 1,850	2.96.1
COTYLEDON COLOR	Yellow	Yellow 6,022	Green 2,001	3.01:1
SEED-COAT COLOR	Gray	Gray 705	White 224	3.15:1
POD SHAPE	Unconstricted	Unconstricted 882	Constricted 299	2.95:1
POD COLOR	Green	Green 428	Yellow 152	2.82:1
POD POSITION	Axial	Axial 651	Terminal 207	3.14:1
STEM LENGTH	Tall	Tall 787	Dwarf 277	2 84:1
★ ★ ★	TOTALS	14,949 Show the dominant	5,010 Show the recessive	2 98 1
SUMMARY	When an F_1 hybrid is selfed	some of the F_2 plants show the DOMINANT TRAIT, and some show the RECESSIVE TRAIT		in the ratio of approximately 3.1

PHILIP GOLDSTEIN

FIG. 5. Results of Mendel's "second set of experiments."

Question 2 What happens to the recessive trait which disappears in the F_1 generation?

For example, when a tall variety was crossed with a short variety, all the F_1 plants were tall. What had happened to the shortness? Was it lost forever, or was it merely hidden temporarily? To find out, Mendel permitted the F_1 hybrids to self-fertilize, and then he made a careful statistical study of the resulting F_2 plants. The results as he reported them are summarized in Fig. 5.

Mendel observed that although the recessive trait disappeared in the F_1 generation, it was not lost, since it reappeared in the next generation. Approximately one fourth of the F_2 plants showed this formerly hidden trait. Furthermore, the recessive trait was in no way altered or contaminated by its close association with the overpowering dominant, for it reappeared in its own form, unchanged in any respect. In fact, when the F_2 plants which showed the recessive trait were permitted to self-fertilize, they continued to produce offspring showing only the recessive trait, for as many generations as the experiment was carried out. Evidently the recessive trait remained pure.

By contrast, the F_2 plants which showed the dominant trait seemed to be of two kinds when they were self-fertilized. There were those that gave rise only to plants showing the dominant trait for as many generations as they were studied. Then there were those which acted like the original F_1 hybrids, in that they produced both dominant and recessive offspring in the ratio of 3:1. Evidently some F_2 dominant plants were *pure* for this trait, whereas other F_2 dominant plants carried the hidden recessive, and therefore were *hybrid*. The entire story of these plants is shown diagrammatically in Fig. 6. Although the diagram shows only flower color, it is typical of the other traits which Mendel studied.

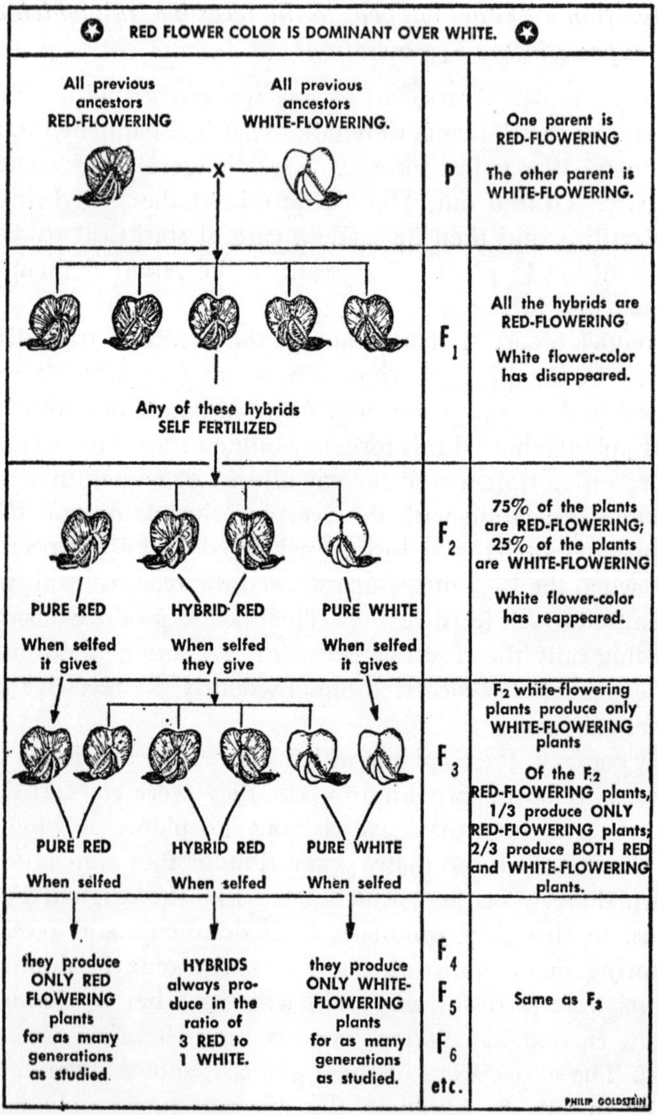

FIG. 6. Inheritance of flower color in Mendel's experiment.

This diagram makes it clear that inheritance in pea plants follows a definite pattern. Mendel expressed this pattern as follows. Let A represent the constant dominant factor, and a the constant recessive factor. Then the hybrid must be Aa, since during self-fertilization it gives rise to plants showing the recessive trait as well as plants showing the dominant trait. The ratio is always $AA + 2Aa + aa$, or 1 pure dominant + 2 hybrids + 1 pure recessive. Mendel pointed out that although the pure dominant (AA) and the hybrids (Aa) look so much alike that they cannot be distinguished, nevertheless they are fundamentally different since the pure (AA) can pass on to its offspring only A, whereas the hybrid (Aa) can pass on to its offspring either A or a.

Up to this point, Mendel had experimented with only one trait at a time. For example, when he dealt with the length of stem, he completely disregraded position of flowers on the stem. Conversely, when he dealt with position of flower, he disregarded length of stem. The fact that he did so helped him succeed in forming the generalizations which no previous worker had reached. Now he was ready to turn to the more complex problem of studying the inheritance of several traits simultaneously. So we reach the third question which Mendel proposed to answer.

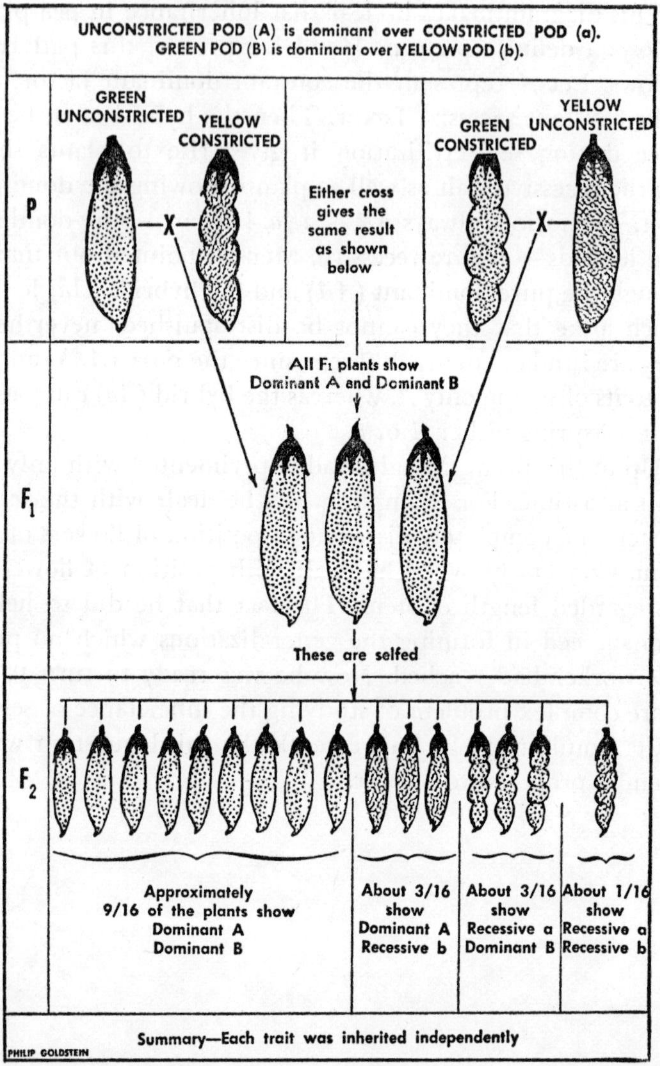

FIG. 7. The inheritance of two traits simultaneously as demonstrated by Mendel's experiment.

Question 3 When several contrasting characteristics are involved in the same cross, are they inherited independently, or are they inherited as a group?

For example, if he crossed a plant having constricted yellow pods with another having unconstricted green pods, what would result? Or if, instead, he crossed a plant having unconstricted yellow pods with one characterized by constricted green pods, how would the results compare? Mendel performed both crosses. The F_1 plants always showed both dominant traits. It made no difference whether both dominants had originally been in one parent, or whether each of the parents had contributed one of the dominant traits. Furthermore, the F_2 derived by self-fertilizing the F_1 plants was also the same in either case. This is shown diagrammatically in Fig. 7. It is noteworthy that pod color is in no way tied to pod shape; rather they are inherited independently of one another. This is evident from the fact that in the F_2 generation, every possible combination of traits appears—constricted green, constricted yellow, unconstricted green, and unconstricted yellow.

Similar experiments were carried out with other traits two or three at a time, and always the same thing happened. In the F_1 generation, only the dominant traits appeared, but in the F_2 generation, the recessives reappeared, giving every possible combination of traits. Evidently each trait is inherited independently of every other trait, and the number of combinations of traits which can appear is limited only by the number of combinations mathematically possible.

Mendel interpreted this as follows: If the result of selfing an organism hybrid for Aa is $(AA + 2Aa + aa)$, and the result of selfing an organism hybrid for Bb is $(BB + 2Bb + bb)$, then the result of selfing dihybrid $AaBb$ is $(AA + 2Aa + aa) \times (BB + 2Bb + bb)$. If a third hybrid trait is introduced, then the result of selfing trihybrid $AaBbCc$ is $(AA + 2Aa + aa) \times (BB + 2Bb + bb) \times (CC + 2Cc + cc)$, etc.

CHAPTER 4

SUMMARY OF MENDEL'S IDEAS ON HEREDITY

As a result of his experimental efforts to answer the questions which he posed for himself, Mendel formulated certain ideas on the method by which heredity operates. These major ideas may be summarized as follows:*

1. Mendel pointed out that characteristics appeared in two alternative forms, such as green or yellow cotyledon color. (Today we call these contrasting pairs *allelomorphs* or *alleles.*)

2. Mendel believed that every trait in an organism was determined by a single pair of hereditary units. He used the letters *A, a, B, b,* etc., to represent them. (Today we call these hereditary determiners *genes.*)

3. Mendel demonstrated that one member of each contrasting pair, the *dominant,* could completely mask the other, or *recessive,* trait in a cross. (We still use the same terms today.)

---

* Sometimes these principles are expressed as the *so-called* "Mendelian Laws of Heredity."

1. The "Law" of Dominance. When organisms pure for opposite traits of a pair are crossed, only one trait of the pair appears in the F_1 generation.

2. The "Law" of Segregation. The characteristic which was hidden in the F_1 is not lost, since it reappears in a definite percentage of the F_2 organisms.

3. The "Law" of Independent Assortment (also known as the "Law" of Unit Characters). Every characteristic is inherited independently of every other characteristic.

However, these principles can hardly be called *laws* because there are as many exceptions to each rule as there are cases that follow the rule.

4. Mendel reasoned that the pairs of determiners which reside together in a hybrid, separate from each other during the formation of reproductive cells, without in any way being contaminated by their previous association with their opposites. As a result, each reproductive cell has only one determiner for each trait. (Today, we use the term *segregation* to describe the separation of the genes of a pair.)

5. Mendel believed that in each generation there is a *recombination* of the determiners carried by the reproductive cells. The combination of determiners from the male parent and from the female parent is dependent purely on chance, and therefore, when there are many offspring, every possible combination of traits is produced. (We still use the term *recombination* to describe this idea.)

Of these five principles, the last two, *segregation* and *recombination,* are really Mendel's distinct contributions, since earlier experimenters had already observed and discussed the others. But the brand-new principles of segregation and recombination effectively changed our entire concept of heredity.

Before Mendel, heredity was considered to be a merging or blending of different bloods. Once blended, the two streams became inseparably merged into one. This is comparable to the mixing of red and blue paint. A new purple product results, from which the original red and blue can never be recovered.

Mendel's contribution was the idea that the original factors which go into the blend can be recovered unchanged. His concept of heredity suggested a group of constant units which remained unaltered and uncontaminated, though shuffled and reshuffled during their transmission from generation to generation.

CHAPTER 5

AFTER MENDEL

MENDEL'S PAPER was read before the Natural History Society in Brünn on February 8 and March 8, 1865. It appeared in print in 1866, and was promptly forgotten. Why this remarkable piece of work was overlooked, though it ranks as one of the outstanding scientific contributions of all time, has never been clearly explained, since the *Proceedings* of the Brünn Society were exchanged with many other scientific societies. However, Darwin's *Origin of Species* had just been published in 1859, and the scientific world was in the midst of a tremendous controversy over Darwinism. It is sometimes stated that the biologists of the day were too busily engrossed in this controversy to notice the work of an obscure Augustinian monk. But this does not seem logical, for, at that very time, the scientific world was hunting high and low for an explanation of heredity as an aid to understanding evolution. This is evident from the fact that dozens of vague and often fantastic theories of heredity were proposed between 1860 and 1890. And all this time, Mendel's carefully worked out explanation remained unknown.

Though Mendel's work remained unknown in this period, experimentation on heredity was being carried out all over the world. Toward the close of the nineteenth century, several men were beginning to develop their own simple laws of heredity just as Mendel had done thirty years before. For example, there was Bateson in England who in 1894, began to stress what he called discontinuous variations, comparable to Mendel's unit characters. In the United States, Spillman actually noted segregation in wheat in 1898. Then suddenly, in 1900, from three different directions, came the announcement that this was "old stuff"; that the laws of heredity had been worked out in 1865. The men who made this announcement (Hugo De Vries in Holland, Karl Correns in Germany, and Eric von Tschermak in Austria) had each carried out his own experiments on heredity prior to reading Mendel's paper. Their data showed that Mendel's laws were not confined to the garden pea but applied to other plants as well.* Within a few years the work of Bateson and Punnett in England, and Cuenot in France, made it clear that the same laws applied to animals also.

So, in 1900, *genetics,* the new science of heredity, was born; and in a few years, this infant science grew by prodigious strides. Credit for much of this growth must go to Thomas Hunt Morgan and the group which worked under his inspirational leadership at Columbia University. Credit must also be accorded to an insignificant little organism, the tiny red-eyed fruit fly (*Drosophila*) which can be seen hovering around bananas or other fruit during almost any part of the year. This fly, with its rapid rate of reproduction (25 generations per year) and its small requirements (hundreds live comfortably in a milk bottle), provided Morgan and his co-workers with ideal material for their experimental work. In fact, it can be stated safely that in the early part of the twentieth century, more was learned about heredity from the study of *Drosophila* than from all other sources put together.

In the space of a few years, thousands of experiments and countless discoveries have been reported. Some of them seem immediately important, others insignificant. Yet who is to

* The research papers of DeVries, Correns, and Tschermak announcing the rediscovery of Mendel's principles are analyzed on pages 148-165 of the book, *TRIUMPHS OF BIOLOGY,* by Philip Goldstein, Doubleday and Co., New York, 1965.

judge their value twenty years, or fifty, or one hundred years from today? Who can judge the direction of future genetic research? However, it must be pointed out in passing that genetics has formed several close partnerships with older branches of science.

For example, genetics leans heavily upon mathematics. Even Mendel used a mathematical approach to his explanation of the mechanism of inheritance. Today, statistics, probability, and mathematical analysis are constantly playing a part in our studies of the inheritance of individual traits or of the traits of whole populations.

Early in its history, genetics teamed up with *cytology,* the study of cell structure. Together they have solved many problems about the mechanics of heredity, and each has provided support for the findings of the other. This union began early in the 1900's and became closer and closer as the years progressed.

Similarly, physics was soon brought into the study of heredity, when X rays, ultraviolet light, atomic energy, and other forms of radiation were used to help genetic research. In 1946 Herman J. Muller was granted the Nobel Prize in Medicine and Physiology for his studies of the effect of X rays on inheritance in *Drosophila.*

A more recent partnership is the one between genetics and chemistry. Under the title of *Biochemical Genetics,* this partnership seeks to understand the effects of heredity on the chemical reactions which are carried on in the bodies of living things. Or, to put it another way, this partnership seeks to explain how genes control the fundamental chemistry of life itself.

The pioneer in this fascinating field of investigation was a British physician named Archibald E. Garrod. Around 1899, Dr. Garrod became interested in a rather rare, but curious human defect (ALKAPTONURIA) in which the urine turned black upon exposure to air. Two facts stood out in his investigations. Alkaptonuria is present at birth. It is often found in more than one child of the same family. These facts seemed to imply that the defect is hereditary. Further research indicated that alkaptonuria resulted from a block in a normal body process.

Garrod believed that there were many such hereditary defects which remained unrecognized. Only those that advertised their presence in some unusual way were recognized. He knew of four such conditions: ALKAPTONURIA, advertised by blackening of the urine; ALBINISM, by absence of pigment in skin, hair and eyes; CYSTINURIA, by formation of kidney stones made of almost pure cystine (an amino acid); PENTOSURIA, by presence of an unusual sugar in the urine.

In 1908 Dr. Garrod published a concise summary of his work under the title *Inborn Errors of Metabolism*. Each of the four conditions he described was caused by a defect (error) in heredity affecting the body chemistry (metabolism). Unfortunately the world was not ready to accept such a radical idea, and Garrod's "Inborn Errors" were completely forgotten.

In the 1930-1940 era, biologists again began to wonder about the hereditary control of chemical body processes. Without knowing anything about Garrod's work, George W. Beadle and Edward L. Tatum set themselves the task of creating chemical defects artificially in the mold *Neurospora*. They succeeded in producing hereditary defects similar to Garrod's inborn errors, and the science of biochemical genetics was "put on the map" to stay. The same methods were then applied to bacteria. Joshua Lederberg used them so effectively that a whole science of bacterial genetics came into being. For these important contributions, George W. Beadle and Edward L. Tatum shared the 1958 Nobel Prize award with Joshua Lederberg.

The medical world was now forced to show some interest

in Garrod's idea of inborn metabolic defects. Many other human defects have since been added to the original list of four. Archibald Garrod has taken his rightful place as "father" of the science of medical genetics.

Establishment of *biochemical, bacterial* and *medical genetics* was only a start. Research soon moved into the area of *dynamic* or *physiological* or *developmental genetics,* as it is variously called. This phase of genetics brings together the geneticist, the embryologist (who studies the development of an organism from the egg), and the physiologist (who is interested in the life functions carried on by the organism). Jointly, they are endeavoring to study the developmental processes by which a gene causes a certain trait to emerge in the embryo and, later, to function in the adult.

More recently emphasis has shifted to genetics on the molecular level. Biologists wanted to locate the very molecules that control heredity, and learn how they transmit traits from generation to generation. Attention was focused on two chemical substances, DNA and RNA. Naturally, there were many attempts to analyze or synthesize these substances. In due time, Dr. Arthur Kornberg discovered an enzyme which could cause a bit of DNA in a test-tube to synthesize more DNA of exactly the same type. At the same time, Dr. Severo Ochoa found a comparable enzyme that synthesized RNA. These two men shared the 1959 Nobel Prize for their discovery.

A group of physicists went one step further with the analysis of DNA. Dr. M. H. F. Wilkins applied X-ray analysis to molecules of DNA, and made certain critical measurements. Based on these measurements, the team of J. D. Watson and F. H. C. Crick constructed a model of the DNA molecule which explains how the molecule can reproduce itself so faithfully, and transmit heredity so exactly. For this brilliant job, Doctors Wilkins, Watson and Crick were awarded the Nobel Prize in 1962.

Evidently DNA molecules send instructions to the cells for the synthesis of various proteins. But the messages must be in some kind of code. While the Watson-Crick model gives a deep insight into the functioning of the DNA molecule, it hardly explains the code. So the burning question of the day became: What is "the code of life"? Can this code be cracked so that the messages transmitted by the DNA can be deciphered? Great progress has been made during the last few years on the problem of deciphering the genetic code.

But what causes a DNA molecule to send instructions to the cells at certain times, and not at other times? What throws the switch to start a DNA molecule working? And what shuts it off again so that it does not go on working forever? This fascinating question was recently attacked by many geneticists in different laboratories all over the world. It seems logical that such a mechanism must exist, or the synthesis of certain substances would go on without ever stopping. This might even be what happens when cancer cells grow in an uncontrolled fashion.

In 1965 three French biologists from the Pasteur Institute (François Jacob, Andre Lwoff, and Jacques Monod) were awarded a Nobel Prize for their discoveries concerning *"genetic control of enzyme and virus synthesis."* In plain English, their studies seem to indicate that there are genes to regulate the working of other genes; there are genes that "turn on" other genes, or "turn them off" again.

CHAPTER 6

THE NUCLEUS CARRIES THE HEREDITY

CELLS ARE the basis of heredity, just as they are the basis of all structure and function in living things. We know that the cells of all organisms are fundamentally similar in structure, consisting of a protective wall surrounding a mass of living material called *protoplasm*. The protoplasm is differentiated into a denser part called the *nucleus,* and a less dense part called the *cytoplasm.* Of these structures, it is the nucleus which is directly associated with heredity.

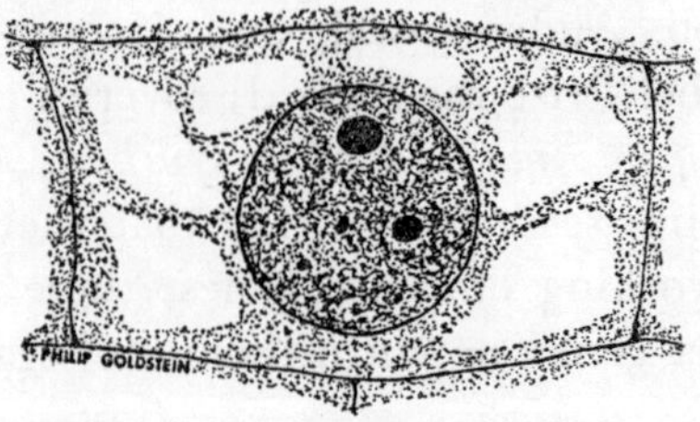

FIG. 8. A cell and its nucleus.

Although the nucleus was observed by several early investigators, it attracted little attention until 1831, when Robert Brown pointed out that nearly every cell had one, and that therefore it must be considered an important part of cell structure. Nevertheless, the function of the nucleus remained obscure until, in 1866, it was suggested that the nucleus might transmit the heredity of the cell. This idea has proved to be correct.

Structurally, the nucleus usually appears as a dense sphere in the protoplasm, more or less centrally located. It is separated from the cytoplasm by a distinct *nuclear membrane,* within which a tangled mass of fine granules and threads can

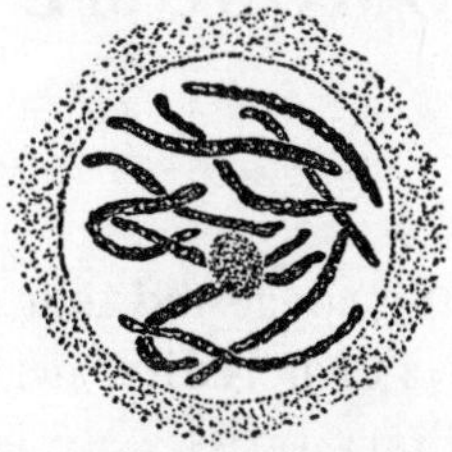

FIG. 9. A nucleus with its chromosomes.

be seen. Because some of these granules and threads readily absorb stains, they are described as *chromatin.* Sometimes, however, the nucleus does not show this characteristic structure. Diagrams of *dividing* cells, drawn in 1848, showed that tiny rodlike structures had been observed within the nucleus in place of the tangled mass of chromatin. In 1875 the conclusion was reached that these rods were actually an important part of the nucleus, perhaps even the carriers of heredity. Later, the name *chromosomes* was applied to these rods.

CHAPTER 7

CHROMOSOMES AND CHROMOSOME NUMBER

SOON, August Weismann suggested that if these chromosomes were the actual carriers of heredity, perhaps the determiners for the various hereditary traits were grouped in single file along their length like a string of beads (see Fig. 10). Of course, all this was theoretical, since nobody had ever observed such an arrangement in those days. As a result of his assumptions, Weismann stated in 1887 that there must be two types of nuclear division.

1. Nuclear division in which every hereditary determiner is split into two equal parts. This could be brought about by the lengthwise splitting of the chromosomes which was reported at about that time (see Fig. 10). As a result, each daughter nucleus would receive exactly the same number and kind of determiners as the mother nucleus. Weismann called this the equal division of the nucleus. Today, we call it *mitosis*. However, our concept of what happens to the chromosomes during mitosis is somewhat different from Weismann's. He believed that each determiner split equally in half during mitotic division. Today, we think that each gene synthesizes an exact copy of itself, which it deposits alongside. Thus another chromosome is produced, identical gene for gene with the original.

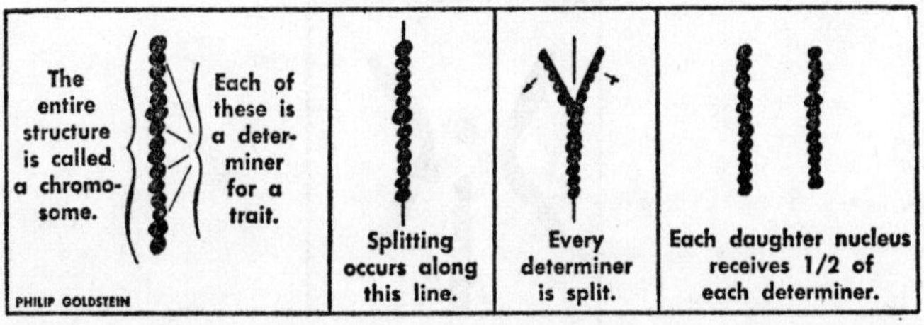

FIG. 10. Weismann's concept of a chromosome and how it splits. (We do not accept this view today.)

2. Nuclear division in which each daughter nucleus receives exactly one half of the total number of hereditary determiners as the mother nucleus. Such a reduction in the number of hereditary determiners is necessary to prevent doubling of the number of determiners each time a male and female cell unite during reproduction. Reduction in number of hereditary determiners could be brought about by the elimination of one half of the chromosomes from the nucleus during the formation of sperms and eggs. Although such behavior had never been observed, Weismann predicted that it would be discovered. He called such division *reduction division,* a name we still use today.

Many of Weismann's guesses have proved to be accurate. We still picture the chromosomes as a string of hereditary determiners, arranged in single file, but we call the determiners *genes.* We know today that chromosomes come in sets, and that, normally, the number of chromosomes in a set is con-

FIG. 11. A set of *Drosophila* chromosomes.

stant for all the organisms of a species. For example, in the famous fruit fly *Drosophila melanogaster*, a set consists of four chromosomes which may be diagramed as in Fig. 11. The number of chromosomes which make up the basic set for any species is called the *monoploid number* (or haploid number) of chromosomes, and every reproductive cell or *gamete* carries one such set of chromosomes in its nucleus.

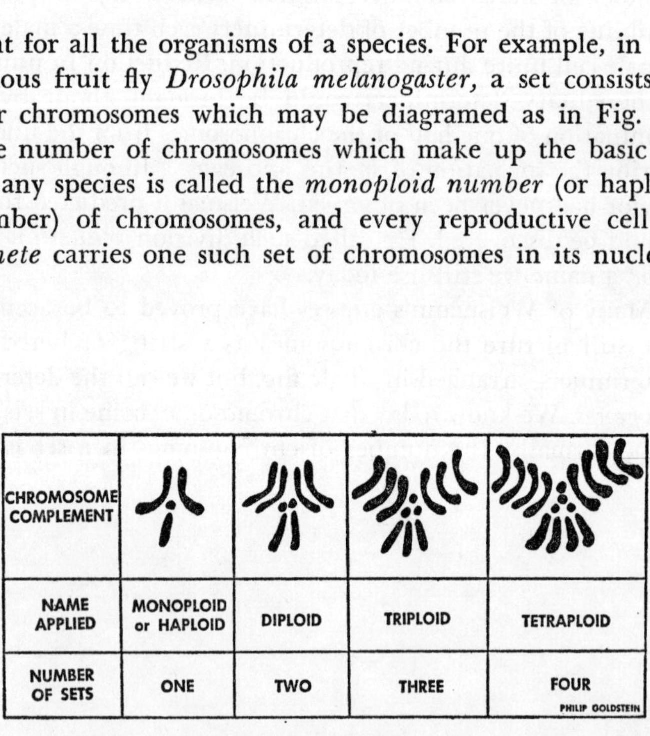

FIG. 12. Polyploidy—Chromosomes come in sets.

When two gametes unite, the resulting *zygote* receives two sets of chromosomes, one from the male gamete, and one from the female gamete, forming a double set of chromosomes called the *diploid number.* If a cell has three sets of chromosomes, such as the cells of corn endosperm, it is said to be *triploid.* If there are four sets of chromosomes, the cell is said to be *tetraploid.* And so forth. This is illustrated in Fig. 12, using *Drosophila* chromosomes as an example. The chromosome numbers characteristic of several other plants and animals are given in Fig. 13.

Name	*Monoploid*	*Diploid*
Fruit fly	4	8
Man	23	46
Corn	10	20
Jimson weed	12	24
Wheat (Emmer group)	14	28
Wheat (common group)	21	42
Pea	7	14
Frog	13	26
Monkey	24	48
Earthworm	16	32
Gypsy Moth	31	62

FIG. 13. A few characteristic chromosome numbers.

CHAPTER 8

HOW THE CHROMOSOME NUMBER OF A SPECIES IS KEPT CONSTANT

AN ORGANISM usually begins life as a fertilized egg formed by the union of an egg and a sperm. From Fig. 14, which illustrates this for *Drosophila,* it is evident that the fertilized

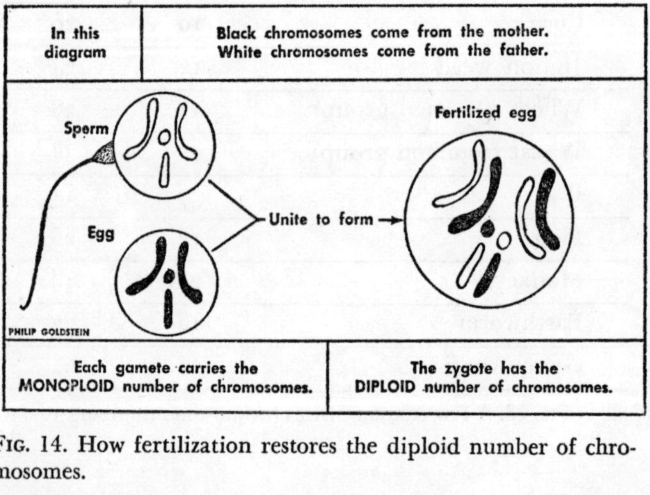

FIG. 14. How fertilization restores the diploid number of chromosomes.

egg received one complete set of chromosomes from its male parent, and a second complete set of chromosomes from its female parent. *Obviously, heredity from both parents is equal.*

The one-celled fertilized egg develops into a many-celled organism by a series of cell divisions. However, each cell division is preceded by a nuclear division, which occurs in a very precise way. At first, the nucleus appears to consist of a tangled network of chromatin, but gradually a specific number of chromosomes emerges from the network. Chromosome counts show that this is always the diploid number for the species. Now these chromosomes begin a strange dance. They are drawn as if by some compelling force to the center of the cell, where every chromosome duplicates itself exactly, particle for particle. The halves seem to repel each other, and, as a result, they move to the opposite sides of the cell, where they gradually lose their identity. Each group of chromosomes becomes a tangled mass of chromatin, and the dance is over. Nuclear division is complete. Soon the cell divides, and now there are two cells, each having a nucleus containing the diploid number of chromosomes. This type of nuclear division is called *mitosis*.

We have seen how the fertilized egg receives its double set of chromosomes, and how the diploid number of chromosomes is maintained as the egg cell divides. All the billions of cells of an organism develop in the same fashion. Each time a cell is ready to divide, the nucleus undergoes mitotic division before cell division occurs. As a result, every cell in the body has the diploid number of chromosomes, just like the fertilized egg. But even more than that, every cell in the body receives exactly the same genes as every other cell, since every gene is duplicated during the process, and identical genes go to each daughter cell.

Soon this organism reaches sexual maturity. It is now ready to produce gametes, or reproductive cells. But it has already been pointed out that the gametes are monoploid, whereas the body cells are diploid. Hence, during the maturation of the gametes, there must be a reduction of chromosome number from two sets (diploid) to one set (monoploid). This is brought about by a type of nuclear division called *reduction division*. This process also begins with a nucleus containing a mass of tangled chromatin. Here too, the dance begins with the emergence of the diploid number of chromosomes from the tangled mass. But this time, each chromosome seeks out a partner with which to dance. Each chromosome carefully picks out and pairs off with the corresponding chromosome from the other set, and they move to the center of the cell together. They cling closely to each other, these *homologous* (corresponding) chromosomes, in an embrace known as *synapsis,* until finally they are pulled apart. Each member of the pair moves away from its mate *without duplication of its genes.* (Note that this statement is not accurate. It is an oversimplification of what really happens. Actually, after the chromosomes go into synapsis, they duplicate themselves and in this way a tetrad, consisting of four chromosomes, is formed.) Thus, as a result, one complete set of chromosomes moves to one side of the cell, and the second complete set of chromosomes moves to the other side of the cell. Again the chromosomes lose their identity, but now, when the cell divides, each daughter cell remains with a nucleus containing only half as many chromosomes as the mother cell. Each daughter cell is monoploid and is ready to develop into a gamete.

Perhaps the contrast between *mitosis* and *reduction division* can best be shown by a series of diagrams such as those presented in Fig. 15. In the diagrams it is assumed that the organism contains two pairs of chromosomes in each body cell.

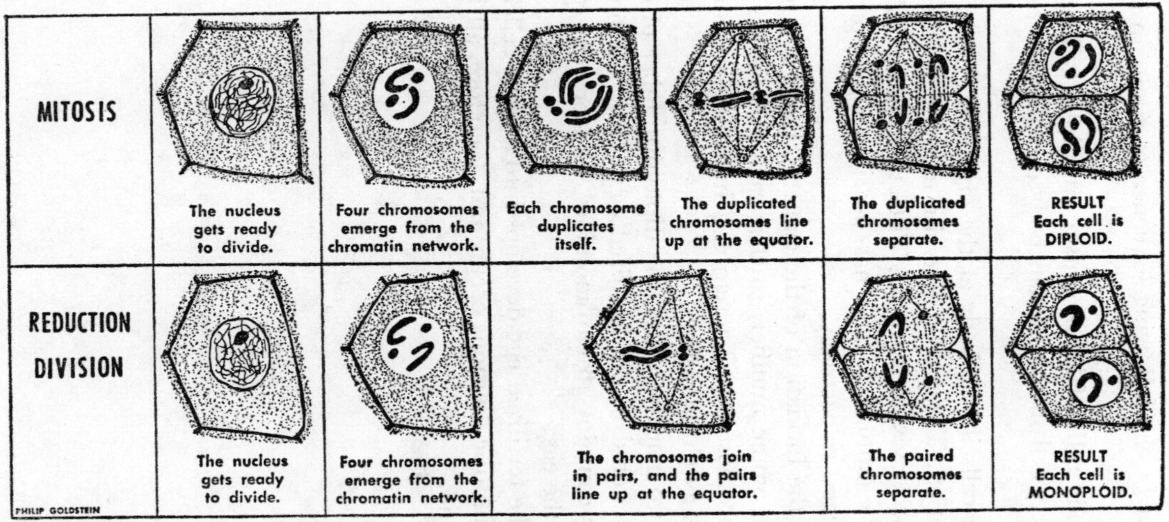

FIG. 15. Mitosis and reduction division compared in a cell having two pairs of chromosomes. (*Note—This is not intended to be an accurate portrayal of what occurs. It is a simplified, diagrammatic version intended only to contrast the results of the two types of nuclear division.*)

To summarize: The chromosome number in a species is kept constant by the following sequence of events in its life cycle.

1. The body cells of an organism are diploid. Thus, for example, every cell in the human body contains 46 chromosomes.

2. Certain cells in the body of the organism are set aside for reproduction. These cells give rise to gametes when the organism reaches sexual maturity. Thus the human male has testes and the human female has ovaries, which produce sperms and eggs, respectively.

3. During the formation of the gametes, there is a reduction in chromosome number, so that each gamete has only one set of chromosomes. Therefore, a human sperm or egg each contains 23 chromosomes.

4. When two gametes unite, the diploid number of chromosomes is restored in the fertilized egg. Thus, the human fertilized egg contains 46 chromosomes, 23 from the sperm and 23 from the egg.

5. When the fertilized egg develops into a many-celled organism only mitotic divisions occur, and, therefore, all body cells are diploid.

This cycle is shown diagrammatically in Fig. 16, where N represents the monoploid number of chromosomes. *While this sequence of events serves to keep the chromosome number constant, it also plays an important part in determining which hereditary traits are passed on.*

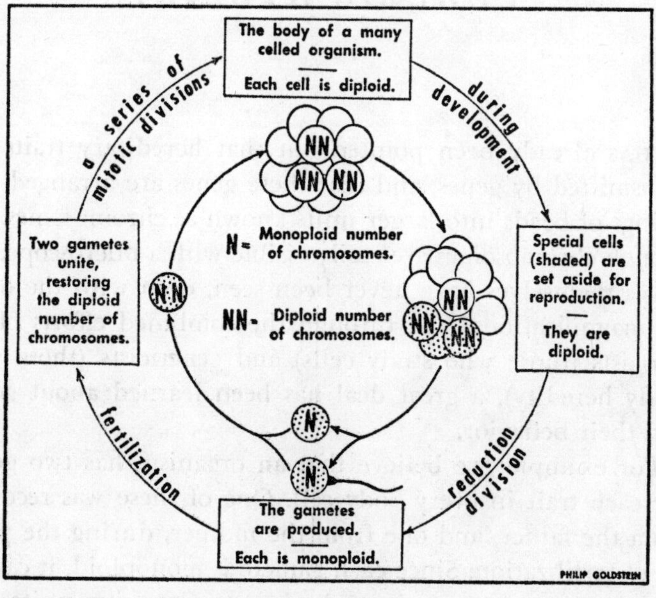

FIG. 16. How the chromosome number is kept constant from one generation to the next in an organism which reproduces sexually.

CHAPTER 9

PURE OR HYBRID?

IT HAS already been pointed out that hereditary traits are transmitted by genes, and that these genes are arranged like strings of beads into larger units known as chromosomes. Although chromosomes are easily visible with a microscope, the genes themselves have never been seen, even with the highest magnifications. Yet, through the combined efforts of cytologists (those who study cells) and geneticists (those who study heredity), a great deal has been learned about genes and their behavior.

For example, we believe that an organism has two genes for each trait in every body cell. One of these was received from the father, and one from the mother, during the process of fertilization. Since each gamete is monoploid, it carries just one gene for a trait, and when two gametes unite, the zygote has two genes for the trait. This gives rise to several possibilities. The two genes which the organism receives may both be identical, as, for example, in the case of a pea plant which has received two genes for tallness, or two genes for shortness. Such an organism is said to be *pure,* or *homozygous,* for this particular trait. On the other hand, the organism in question may have received a gene for tall from one parent and a gene for short from the other parent, so that it has two opposite genes for height. Such an organism is said to be *hybrid,* or *heterozygous,* for this trait.

We know further that, during reduction division, the chromosomes of a pair separate. Obviously, the two genes for a trait must also separate, each going to a different gamete. As a result, the pure tall pea plant produces a large number of gametes, each of which carries one gene for tallness. Similarly, the pure short pea plant produces many gametes, each carrying one gene for shortness. However, the hybrid produces two types of gametes in equal numbers. Half of them carry the gene for shortness, and the other half carry the gene for tallness.

All this is represented diagrammatically in Fig. 17. In the diagram, it is assumed that *T* represents a gene for tallness,

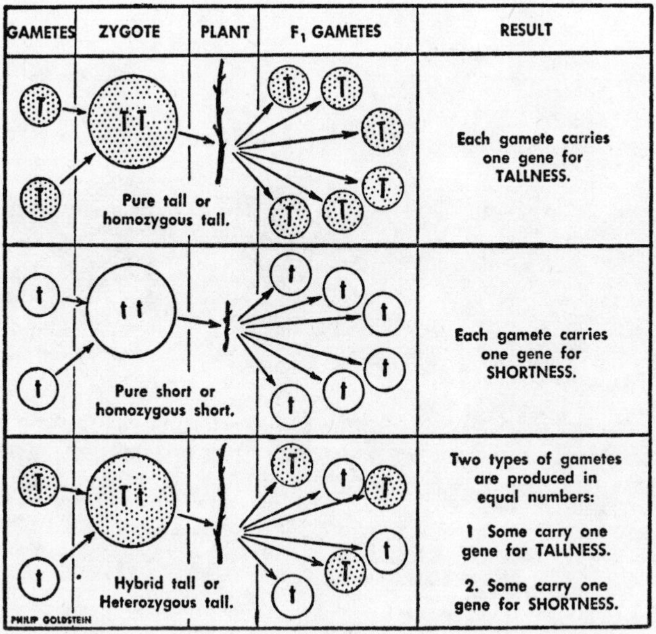

FIG. 17. The gametes produced by pure tall, pure short, and hybrid tall pea plants.

and *t* represents a gene for shortness. It must be understood that there are many other genes in each cell diagramed, but that, for the present, we are disregarding them.

A clear understanding of these simple facts immediately throws light on some of Mendel's results. *The Mendelian ratios emerge as meaningful, understandable outcomes of normal chromosome behavior,* so that what Mendel discovered experimentally, can be explained in terms of the science of cytology. This will be clarified in the sections which follow.

CHAPTER 10

DOMINANCE—COMPLETE OR INCOMPLETE

MENDEL FOUND that when he mated pea plants showing opposite characteristics of a pair, all the F_1 plants showed only one characteristic of the pair. For example, in a tall × short cross, all the children were tall.

Why all the F_1 children must be identical can be explained genetically somewhat as follows. The tall parent is pure for tallness, and so every cell in its body has two genes for tallness. When this tall plant produces gametes, reduction division separates the two genes for tallness, and each gamete receives one of them. As a result, each gamete from the tall plant carries one gene for tallness, and in the same way, each gamete from the short plant carries one gene for shortness. When the gametes from the tall parent unite with those from the short parent, only one result is possible, for each zygote receives one gene for tallness and one gene for shortness. All the F_1 plants *must* be hybrids, with the genetic make-up *Tt*. This can be visualized more clearly by reference to Fig. 18.

Although this explains why all the plants are genetically similar, it does *not* explain why they are all tall. Why are they not all short instead? Or all intermediate? In other words, why tall is dominant over short has not been explained.

At the present writing, a really good explanation is impossible. We do not yet know enough about how genes act to be able to explain why one trait of a pair can overshadow the

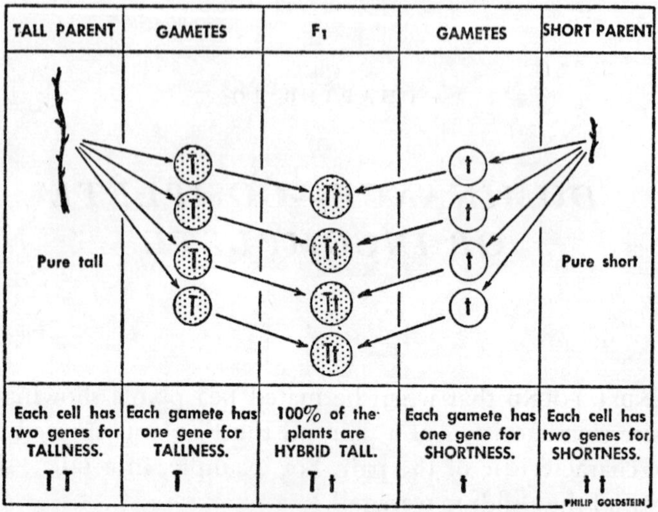

FIG. 18. The result of crossing tall and short pea plants.

other. We can merely say that one gene for tallness is as potent as two genes for tallness in determining the height of a pea plant. For this reason, the pure tall (*TT*) and the hybrid tall (*Tt*) *both appear identical, even though they are genetically different.* This situation makes it necessary to introduce two additional terms—*genotype,* which refers to genetic make-up, and *phenotype,* which refers to appearance regardless of genetic make-up. Thus a pure tall pea plant and a hybrid tall pea plant are phenotypically similar but genotypically different.

To heighten the mystery of dominance, it is interesting to note that because a trait is dominant over its allelomorph in one species, it does not follow that the same will be true in a second species. For example, black is dominant over white in rabbits and guinea pigs. But on the contrary, white is dominant over black in the Leghorn chicken. And to make matters more confusing, in another type of chicken, the Andalusian fowl, the offspring of a black × white cross have a mixture of black and white in their feathers, which presents a slate-blue appearance. In the last case, neither characteristic of the pair is able to overshadow the other, and instead, a blending of the two parental traits results. This situation is known as *incomplete dominance* or *blending inheritance.*

Similarly, the fact that one trait in an organism is completely dominant over its allele does not mean that every trait in that organism will illustrate complete dominance. For example, in man, dark hair color is completely dominant over light hair color, and dark eye color is completely dominant over light eye color. But skin color shows incomplete dominance, so that a negro-white marriage gives rise to mulatto children.

So it is evident that the explanation for dominance is a mystery. Yet, in a few isolated cases, the basis for dominance has been satisfactorily explained. In these cases, few and far between though they are, we *do* understand why one trait can overshadow its opposite. Several of these cases will now be described.

A. Why Normal Body Size Is Dominant Over Dwarf in Mice

Some years ago, a strain of dwarf mice was discovered. When these mice were mated among themselves, only dwarfs were produced, so that evidently dwarfness was a hereditary trait. When a normal mouse was mated with one of these dwarfs, the F_1 offspring were all normal, showing that normal was dominant over dwarf. Anatomical studies of both dwarf and normal strains revealed a marked difference in the size of the pituitary gland. It is known that the pituitary gland produces a hormone which controls growth, and therefore it was suspected that in the dwarf mice an insufficient quantity of this hormone was produced. To test this, normal pituitary tissue was implanted into some baby dwarf mice, and it was found that they grew to normal size, showing this hypothesis to be correct.

In this case, the gene for normal (*N*) is in some way tied up with normal functioning of the pituitary gland, so that in the presence of gene *N*, growth hormone is produced. Since gene *N* is present in mice of genetic make-up *NN* or *Nn*, both these types will be normal in size. However, dwarf mice (*nn*) lack the *N* gene completely, and so they are deficient in growth hormone.

Here then, dominance is caused by the fact that even one gene for the normal (*Nn*) brings about the production of enough growth hormone to stimulate normal growth. The presence of a second gene for normal (*NN*) is merely insurance that growth will be normal. Fig. 19 summarizes this situation.

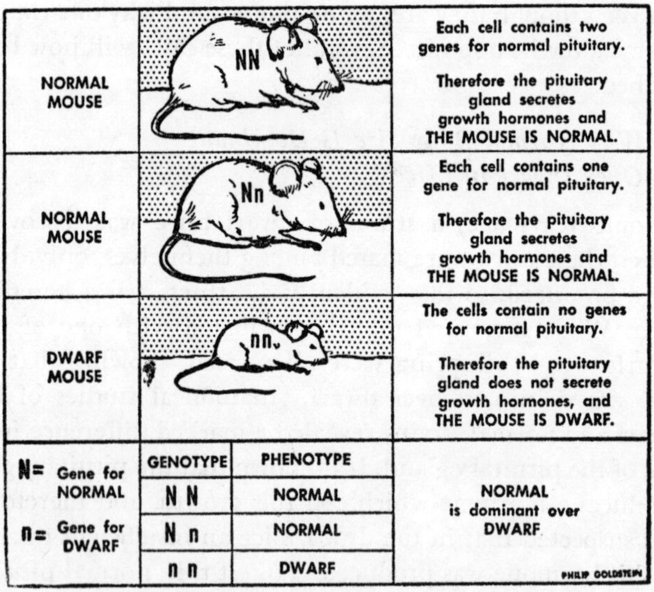

FIG. 19. Why normal is dominant over dwarf in mice.

B. Why Black Is Dominant Over White in Rabbits

A second case in which the reason for dominance has been established deals with color in the rabbit. When a pure black rabbit is crossed with a pure white, all the offspring are black. Hence black is the dominant trait. But why?

Black rabbits are black because the pigment *melanin* is present in the hair. By contrast, white rabbits lack this pigment. However, chemists showed that melanin could be produced in the laboratory by allowing the enzyme *tyrosinase* to act on the amino acid *tyrosine* in a test tube. This raised an interesting question. Both black and white rabbits eat the same food. This food contains the amino acid tyrosine. Yet only black rabbits produce melanin. Could it be that black rabbits have the power to produce the enzyme tyrosinase, while white rabbits cannot?

The question was answered when it was found that an extract made from the skin of a black rabbit gave an active tyrosinase solution, while a similar extract from the skin of a white rabbit was negative for tyrosinase. Evidently then, the gene for black is really tied up with the ability to produce the enzyme, or perhaps it is the enzyme itself. As long as the rabbit has at least one gene for the production of tyrosinase, it is able to convert tyrosine to melanin, and therefore

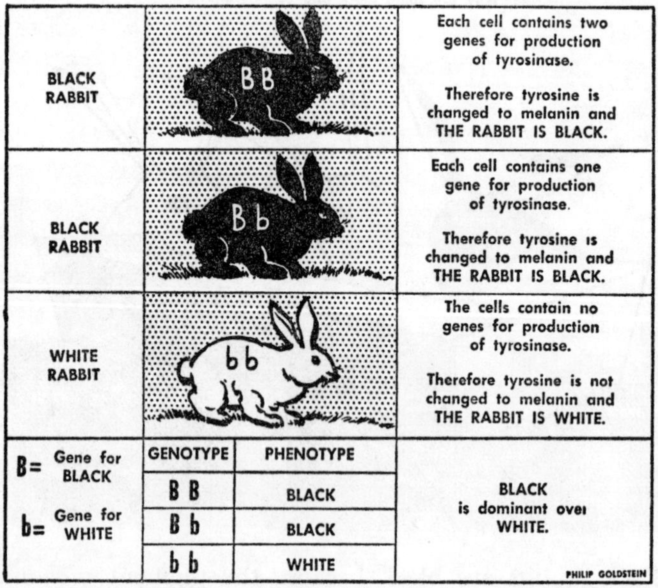

FIG. 20. Why black is dominant over white in rabbits.

it will be black. If it has no genes for the production of the enzyme, it cannot change tyrosine to melanin, and it is white. Fig. 20 explains diagrammatically why black is dominant over white in rabbits.

C. The Case of the Urine That Turned Black

Over three and a half centuries ago, it was noted that the urine of certain people turned black upon exposure to air.

The reason for this was discovered in 1858, when it was learned that the urine of these people contained a substance called alkapton which was totally absent from the urine of normal people. This substance caused the urine to blacken. Soon after 1900, two English pioneers in the field of genetics showed that this defect was inherited as a simple recessive trait, normal being dominant over blackening of urine.* Then, in 1914, an enzyme which destroyed alkapton was found in the blood of normal people. However, people suffering from the defect did not have this enzyme in their blood, and therefore their kidneys excreted alkapton in the urine.

Here lay the clue to the dominance of the normal condition. Evidently, the gene for normal causes the enzyme to be produced. Therefore, whether there is one gene for normal or two genes for normal makes no difference. In either case enough enzyme is produced to destroy the alkapton. It is only in the absence of the normal gene that no enzyme is produced, so that alkapton is not destroyed, and the urine turns black. Therefore normal must be dominant.

D. The Case of the Nonfermenting Yeasts

While most yeasts efficiently ferment the common simple sugars (monosaccharides), they differ in their ability to ferment complex sugars (polysaccharides). The ability to ferment a given polysaccharide seems to depend on whether or not the yeast can produce the enzyme necessary to convert this particular polysaccharide to simple sugars. A yeast lacking the specific enzyme cannot ferment a given polysaccha-

---

* One of these pioneers was a British physician named Archibald E. Garrod who began work on this unusual kind of urine in 1899. A summary of his research is presented on pages 208-220 of the book *TRIUMPHS OF BIOLOGY* by Philip Goldstein, Doubleday and Co., New York, 1965. A more complete description of the work is contained in Dr. Garrod's classic little book entitled *Inborn Errors of Metabolism,* Frowde, Hodder, & Stoughton, London, 1st ed., 1909.

ride, while a yeast which produces the appropriate enzyme can ferment that sugar. The ability to ferment a specific sugar, or the absence of this ability, is inherited.

When it was discovered that yeasts conjugate, crosses were attempted between yeasts which could ferment a given sugar and yeasts which could not. In every case where such a cross was made, fermenting ability proved to be dominant. Evidently the gene for fermenting ability causes the production of an appropriate enzyme. Whether one such gene is present, or two, makes no difference, since in either case the enzyme is produced. A nonfermenter is produced only when no genes for this enzyme are present.

Therefore, fermenting ability must always be dominant over the absence of this ability.

CHAPTER 11

CONSERVATION OF THE GENE—SEGREGATION AND RECOMBINATION

OF ALL the contributions Mendel made to the knowledge of heredity, only the idea of segregation and recombination has stood, without modification, the test of time. We know many exceptions to the Mendelian ideas of dominance, independent assortment, single factors, etc., but under normal circumstances there are *no* exceptions to the idea of segregation.

Essentially, segregation implies that recessives may reappear unchanged after being hidden for one or more generations. For example, although the short trait disappears in the F_1 when a pure tall pea plant is crossed with a pure short, it is not lost, since it reappears in the F_2. Evidently the shortness was in no way affected by its residence within the tall hybrid of the F_1 generation. There are two problems here, both of which can easily be explained on a genetic basis.

The first problem may be stated as follows: "Why does the recessive gene remain uncontaminated by its close association with the dominant gene in the hybrid organism?" For the answer, we must go back to our knowledge of cell structure. Remember that the hybrid has received one gene for the dominant and one gene for the recessive. At no time do these genes fuse, or merge, or blend. At all times, they remain discrete entities, each on its own chromosome. It just happens that for some unknown reason, the action of the dominant gene can mask the action of the recessive gene. But the dominant gene has no effect whatsoever on the recessive gene itself. Therefore, the recessive gene remains unchanged, even though it is in the same cell with its dominant allelomorph, and even though it may remain hidden for many generations.

The second problem follows directly from the first. "Assuming that the recessive trait did remain uncontaminated while it was in the hybrid, why does it reappear in subsequent generations?" Here again there is a simple genetic explanation. In each cell of the hybrid, there is one gene for the dominant and one gene for the recessive. When gametes are formed, these two genes are completely separated from each other during the process of reduction division. As a result, every hybrid produces two types of gametes in equal numbers (see Fig. 17). Some of the gametes carry the gene for the

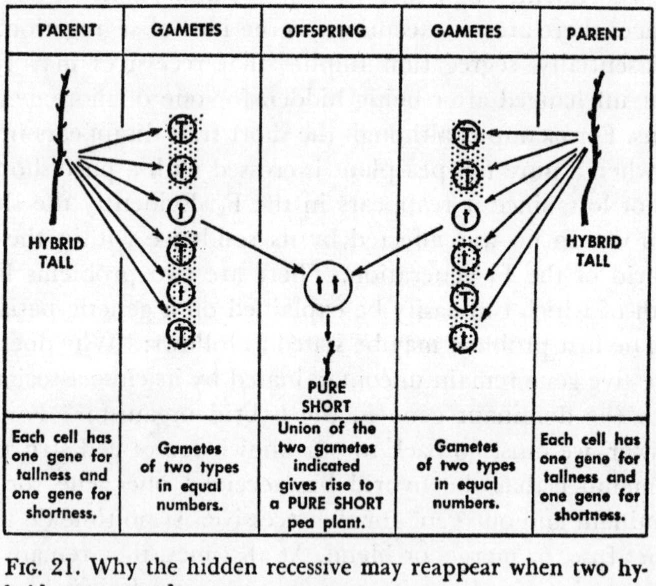

FIG. 21. Why the hidden recessive may reappear when two hybrids are crossed.

recessive. When two hybrids are crossed, the union of two gametes, each carrying the recessive, is certainly possible, and such a fusion gives rise to an organism showing the recessive trait. How this occurs is shown clearly in Fig. 21.

CHAPTER 12

THE MENDELIAN RATIOS AND THE LAW OF CHANCE

THIS IS a good time to consider the so-called "Mendelian ratios" (the ratios which result from various matings). Why, for example, does the mating of two hybrids usually yield offspring of which 75 per cent show the dominant trait, and 25 per cent show the recessive trait? The reason is easy to understand, if we follow the reproductive process clearly. Basically, it depends on the fact that reduction division causes the two genes of a pair to segregate from one another, and go into different gametes.

Let us assume that a red marble represents a gamete carrying a gene for red, and a white marble represents a gamete carrying a gene for white. (A gamete could not possibly have both a gene for red and a gene for white because reduction division separates the two genes of the pair during the formation of the gametes.) A hybrid red organism ready to reproduce might be represented as a jar containing red and white marbles in equal numbers. If we have two such jars, one may be used to represent a female, and the other a male. Now if a *blind* man were to draw one "sperm" from the male jar, and one "egg" from the female jar, this might be an approximation of what happens during fertilization of the egg. What color genes would the fertilized egg receive? Obviously, it is impossible to say definitely what the result would be, but there are only four possibilities, as indicated in Fig. 22.

If only 1 draw is made, the outcome cannot be predicted since the element of chance is involved. Likewise, if 4 draws are made, it is unlikely that we will get one of each type indicated in Fig. 22. But if 400 draws are made, our chances of

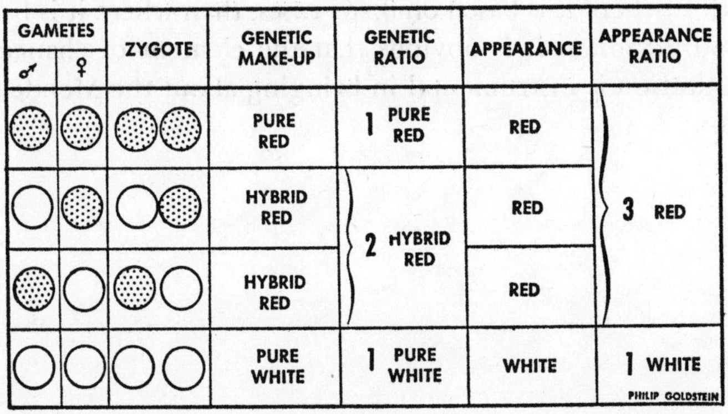

FIG. 22. Explanation of the 3:1 ratio. Red (shaded) is dominant over white.

approaching the ratios indicated in the diagram improve considerably, and if 4,000 draws are made, we are apt to come pretty close to a perfect ratio.

In living things the situation is quite analogous to the one just described. Just as chance determined whether the blind man would draw a red or white marble, so chance determines which sperm will unite with which egg. Therefore, in actuality, the ratio which results from the mating of two hybrids is quite similar to that which results from drawing marbles out of a jar. Here too, it is impossible to predict the appearance of a single offspring. But the larger the number of offspring produced, the closer does the actual ratio approach the theoretical ratio. In this connection it is worth re-examining Mendel's F_2 results (see Fig. 5). To cite only two examples in each of which the theoretical ratio is 3:1:

1. Cotyledon color—a count of 8,201 seeds gave a ratio of 3.01 yellow to 1 green.
2. Pod color —a count of 580 plants gave a ratio of 2.82 green to 1 yellow.

Note how much closer the actual ratio approaches the theoretical where it is based on 8,201 cases, than where it is based on only 580. So it is obvious that the element of chance in fertilization is instrumental in bringing about the Mendelian ratios.

CHAPTER 13

GENES THAT KILL AND MODIFIED MENDELIAN RATIOS

FROM WHAT has just been said, it is obvious that the element of chance in fertilization is instrumental in bringing about the Mendelian ratios. Of course, the actual ratio may be modified by different conditions in different organisms, so that what is actually the same result, may look quite different at first glance. For example, in Fig. 22 we assumed that dominance was complete, and, therefore, the appearance ratio which resulted was 3 red to 1 white. Suppose now, that in another organism, neither red nor white is dominant, but instead there is a blending of red and white to give pink. This occurs in the snapdragon, the four-o'clock, and other organisms. Here the 3:1 ratio is modified to 1 red to 2 pink to 1 white, as Fig. 23 indicates. Note that despite the different appearance ratio, the actual genetic result is identical, namely,

GAMETES		ZYGOTE	DOMINANCE COMPLETE.		DOMINANCE INCOMPLETE.		PURE RED LETHAL.		PURE WHITE LETHAL.	
♂	♀									
●	●	● ●	red	3 red	red	1 red		dies	red	3 red
●	○	● ○	red		pink	2 pink	red	2 red	red	
○	●	○ ●	red		pink		red		red	
○	○	○ ○	white	1 white	white	1 white	white	1 white		dies

FIG. 23. Modifications of the 3:1 Ratio.

1 pure for red, 1 pure for white, and 2 hybrids, each having a gene for red and a gene for white.

Another modification of the 3:1 ratio is also possible. Let us assume that every zygote which receives two genes for red dies (for reasons unknown to us at the present time). The ratio now becomes 2 red to 1 white, as Fig. 23 shows. Genes which kill are called *lethals,* and they are quite common. Over 50 lethals are known in *Drosophila* alone, some of which kill the egg, some the larvae, some the pupae, and some the adult. To illustrate the last of these, it may be pointed out that flies which are pure for vestigial wings live only about half as long as normal-winged flies under controlled laboratory conditions. Evidently the gene for vestigial has a weakening effect which leads to early death. In mice, yellow is dominant over black, but no pure yellow mice exist since the presence of two genes for yellow kills the embryo at an early stage of its development. In corn and other plants, there is a gene called *albino* which inhibits chlorophyll development. Obviously, those plants which are pure for albino soon die, since they are unable to manufacture their own food.

CHAPTER 14

WHY TRAITS ARE INHERITED INDEPENDENTLY—INDEPENDENT ASSORTMENT

HOW THE ELEMENT of chance involved in fertilization helps determine the results of a mating was explained in an earlier section. Another opportunity for chance to enter the picture arises during the formation of gametes if two or more genes are considered simultaneously. This fact clarifies the genetic basis for Mendel's "Law" of Independent Assortment, which states that every trait is inherited independently of every other trait.

Let us assume that gametes are being formed in a pea plant which is hybrid for two traits—height and flower color. A cell from such a plant is diagramed in Fig. 24. It will be recalled that during the reduction division which accompanies the formation of gametes, the chromosomes line up in pairs, and then the pairs separate so that each gamete receives one member of every pair. Reference to Fig. 24 makes it evident that there are two ways in which these particular chromosomes might line up.

1. *Shaded chromosome No. 2 might find itself on the same side of the cell as shaded chromosome No. 1. As a result, the two shaded chromosomes would go together into one gamete, and the two unshaded chromosomes would go together into a second gamete.*

2. *Shaded chromosome No. 2 might find itself on the same side of the cell as unshaded chromosome No. 1. Hence they would go together into one gamete, while unshaded No. 2 and shaded No. 1 would go together into the second gamete.*

Since reduction division is going on in many cells at once, both of these arrangements actually occur, and, as a result, the dihybrid produces four different genetic types of gametes in approximately equal numbers.

Of course, the greater the number of genes involved, the greater will be the genetic variation of the gametes. Thus, an organism hybrid for 1 trait (Aa) produces 2 types of gametes, A and a; a dihybrid produces 4 types of gametes, AB, Ab, aB, ab; a trihybrid produces 8 types of gametes, etc. In general, the number of types of gametes produced by an organism equals 2^n where n is the number of traits for which the organism is hybrid. Thus, an organism hybrid for only 10 traits can produce over 1,000 kinds of gametes, and an organism hybrid for 20 traits can produce over 1,000,000 types of gametes. Since organisms are actually hybrid for hundreds of different traits, it is evident that the possibilities for genetic variation are practically boundless, and that the combination of genes carried by any particular gamete is entirely dependent on chance.

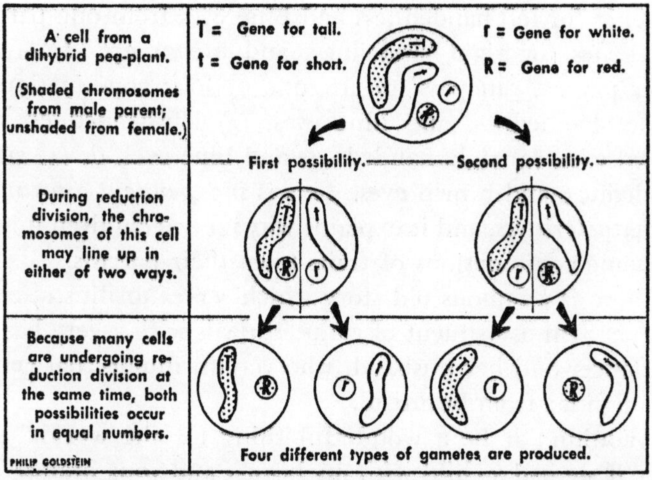

FIG. 24. Why a dihybrid produces gametes of four different types.

Why, then, is every trait inherited independently of every other trait? Simply because there is nothing fixed about the way in which the chromosomes line up during reduction division. Since sheer chance determines how the chromosomes line up in any cell, it is evident that chance and chance alone determines what gene combination any particular gamete will carry. And since there are many cells undergoing reduction division simultaneously, all the possible genetic types will be produced in about equal numbers. It is noteworthy that the chromosomes contributed by a parent to one of its offspring need not necessarily remain together when this offspring produces gametes of its own. Thus a child that inherits genes for left-handedness and blue eyes from one parent, and genes for right-handedness and brown eyes from the other parent, can pass on any one of four combinations—(1) left-handedness and blue eyes, (2) left-handedness and brown eyes, (3) right-handedness and blue eyes, or (4) right-handedness and brown eyes. This is *independent assortment of characteristics,* and it explains why no two children inherit the same combinations of traits from their parents.

There is a famous old story which serves to illustrate the independent assortment of traits. It deals with a very beautiful dancer and her husband, who was an intellectual genius although far from beautiful.

"Wouldn't it be a wonderful thing for the world," said she, "if we had a child with my beauty and your brains?"

"Yes, it would be wonderful indeed," replied her husband. "But can you imagine what a tragedy it would be if the child had *my* beauty and *your* brains!"

That is independent assortment. Even a genius cannot control the combination of traits which his children inherit. It is entirely dependent on chance.

CHAPTER 15

HOW TO SOLVE PROBLEMS

(This section may be disregarded if the reader is not interested in problem solving.)

SINCE CHANCE plays such an important part in determining the results of a mating, it is customary to work out the theoretical ratio by some mechanical means which takes into account all the possible combinations. There are many different methods, only three of which will be described here.

A. Mendel's Algebraic Method—for Mathematicians

This method was mentioned when Mendel's work was described. Considering only one trait at a time, an organism may be pure for the dominant, AA or (A + A); or hybrid, Aa or (A + a); or pure for the recessive, aa or (a + a). The theoretical results may be obtained by multiplying together (as in algebra) the two organisms being mated, thus:

(A + a) ×(a + a)	(A + a) ×(A + a)
Aa + aa Aa + aa	AA + Aa Aa + aa
2Aa + 2aa	AA + 2Aa + aa
When a hybrid is crossed with a pure recessive, the ratio which results is 50% hybrid 50% pure recessive	When two hybrids are crossed, the ratio is 25% pure dominant 50% hybrid 25% pure recessive

A mating involving more than one trait is worked out as in the following problem. "What is the result of self-fertilizing a pea plant which is hybrid tall and hybrid for red flowers?"

$$TtRr \times TtRr \rightarrow ?$$

First work out the results for each trait alone

$$\begin{array}{c} T + t \\ T + t \\ \hline TT + 2Tt + tt \end{array} \qquad \begin{array}{c} R + r \\ R + r \\ \hline RR + 2Rr + rr \end{array}$$

Then multiply the products

$$\begin{array}{c} TT + 2Tt + tt \\ RR + 2Rr + rr \\ \hline \end{array}$$

It seems quite clear that this method becomes rather complicated when two or more traits are involved. As a result, a more efficient method was devised.

B. The Checkerboard Method or Punnett Square—for Novices

The use of the Punnett square may best be illustrated by solving an actual problem. "What is the expected result when two hybrid black guinea pigs are crossed?"

Explanation: Each letter represents a color gene. The male parent produces two genetic types of gametes, B and b, in equal numbers. These are put along one side of the square. The female parent also produces gametes of two types, B and b, and these are put along the other side of the square. Each box in the square represents a possible offspring of the mating. For example, box 1 represents a pure black offspring resulting from the fertilization of egg B by sperm B. Similarly box 2 represents the fertilization of egg B by sperm b, resulting in a hybrid black offspring. The ratios are obtained after all the boxes are filled in, by counting the number of offspring of each type.

Problems involving more than one pair of traits are solved in exactly the same manner, except that the checkerboard may have a greater number of squares. The only difficulty

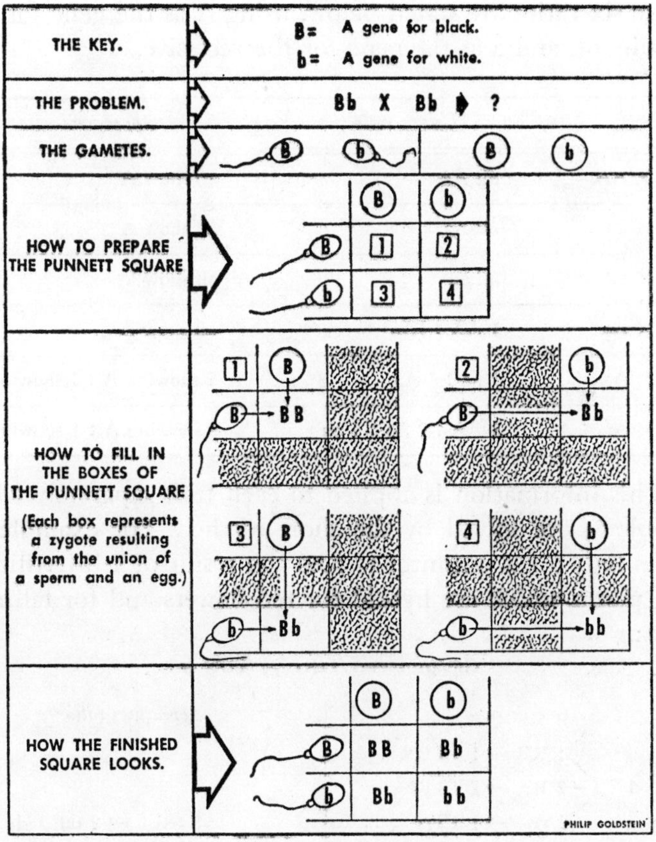

which may arise relates to the gametes. It is essential that all the possible gene combinations be represented in the gametes in their proper proportion (see pp. 57-60 and Fig. 26, p. 68).

C. A Short Method—for Experts

For those who really know genetics, a very simple method of solving complicated problems is available, based on the fact that every trait is inherited independently of every other trait. Actually, to use this method, one need only know the six ratios involving one pair of characteristics. Therefore, these six ratios are stated below, using A as the gene for the dominant, and a as the gene for the recessive.

Parents	*Genetic ratio*	*Appearance ratio*
AA × AA	all AA	all show A
AA × aa	all Aa	all show A
aa × aa	all aa	all show a
AA × Aa	1 AA : 1Aa	all show A
aa × Aa	1 aa : 1 Aa	1 showing A : 1 showing a
Aa × Aa	1 AA : 2 Aa : 1 aa	3 showing A : 1 showing a

This information is applied to each trait separately when problems are solved by the short method. For example, it takes only a few minutes to find the result of self-fertilizing pea plants which are hybrid for red flowers and for tallness, thus:

The problem: TtRr × TtRr → ?

Genetic ratio

1 TT — 1 RR → 1 TTRR
1 TT — 2 Rr → 2 TTRr
1 TT — 1 rr → 1 TTrr

2 Tt — 1 RR → 2 TtRR
2 Tt — 2 Rr → 4 TtRr
2 Tt — 1 rr → 2 Ttrr

1 tt — 1 RR → 1 ttRR
1 tt — 2 Rr → 2 ttRr
1 tt — 1 rr → 1 ttrr

Appearance ratio

3 Tall — 3 red → 9 tall red
3 Tall — 1 white → 3 tall white

1 Short — 3 red → 3 short red
1 Short — 1 white → 1 short white

Explanation (appearance ratio): Consider height alone. Crossing two hybrid tall organisms gives a ratio of 3 tall to 1 short. But among the talls, for every 3 having red flowers there will be 1 with white flowers. The same will also be true of the shorts. Therefore, the complete ratio will be 9 tall red to 3 tall white to 3 short red to 1 short white. The same procedure is followed, with appropriate modifications, to find the genetic ratio.

To show how simple this method is, let us find the appearance ratio in a mating involving six traits.

AABbCCddeeFf × AABbccDdeeFF → ?

To find the ratio of appearance, we proceed as follows:

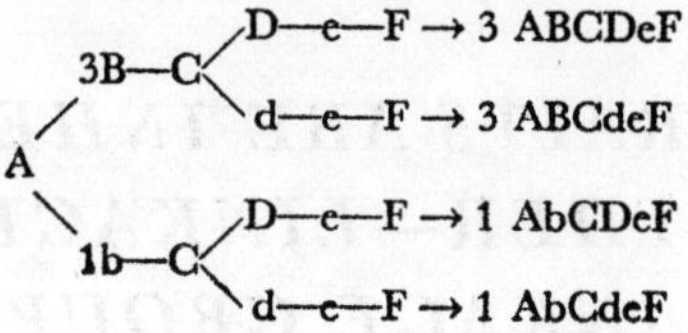

CHAPTER 16

WHY TRAITS ARE INHERITED TOGETHER—LINKAGE AND LINKAGE GROUPS

EVEN MENDEL was aware that his "Law" of Independent Assortment did not always hold true. For example, in his pea plants he found that every plant which had red flowers also had seeds with gray seed coats, whereas every plant that had white flowers had seeds with white seed coats. If flower color and seed-coat color are inherited independently, some plants with red flowers and white seed coats should appear, and some with white flowers and gray seed coats. Yet these types did not appear, and Mendel noted it in his paper without attempting to explain. Today, however, we understand why traits sometimes stick together in this way.

Although an organism may have hundreds or thousands of heritable traits, it is limited to a relatively small number of chromosomes (see Fig. 13). It is evident, therefore, that each chromosome must carry many genes. Since all the genes on one chromosome act as though they are strung together on a common strand, they must travel together during reduction division, and therefore they must be inherited together rather than independently. Genes which tend to travel together on the same chromosome are said to be *linked*, and the condition is known as *linkage*. How linkage affects heredity is explained in Figs. 25 and 26.

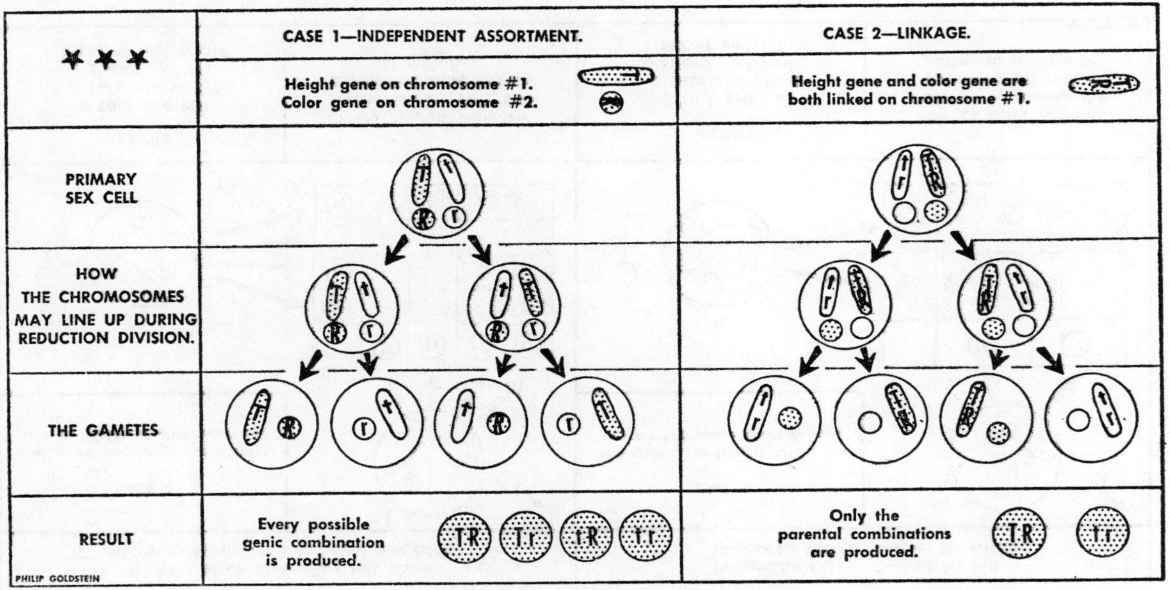

FIG. 25. Gamete formation where linkage exists, compared to gamete formation where there is no linkage.

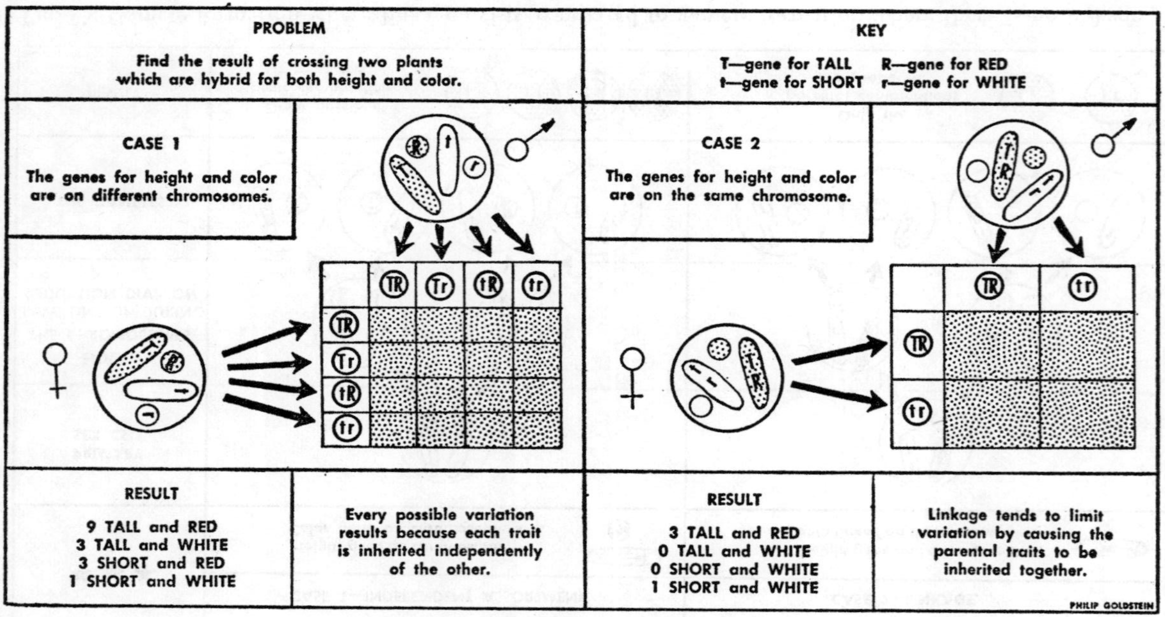

FIG. 26. How linkage affects variation.

Consideration of these facts seems to indicate that traits are not inherited independently but, rather, in groups. All the genes which are linked on one chromosome travel as a group from one generation to the next. Such a group of traits, carried by a single chromosome, comprises what is called a *linkage group*. It is obvious that if this idea of linkage groups is to remain acceptable, the number of linkage groups of traits found in any species must agree with the number of chromosome pairs. Thus in corn there are 10 pairs of chromosomes, and 10 linkage groups are known. Similarly in *Drosophila* where there are only 4 pairs of chromosomes, only 4 groups of linked traits are known. It is interesting to note that in *Drosophila* there is a fairly close correspondence between the length of the chromosomes and the number of traits in the linkage group associated with each chromosome. Thus, if the ratio of length is approximately 7:12:10:1, the number of traits in the corresponding groups are in the ratio of approximately 8:12:9:1. Since man has 23 pairs of chromosomes, there must be 23 human linkage groups.

CHAPTER 17

MAPPING THE CHROMOSOMES—CROSSING OVER

THE PROBLEM of linkage is somewhat complicated by the fact that in some cases the linkage between two traits breaks, owing to the process known as *crossing over* between the chromosomes of a pair. During reduction division, the chromosomes of a pair come together and twine about each other. This is known as *synapsis.* The interesting thing about it is that there seems to be a mutual attraction between the two genes for the same trait, so that when the two chromosomes synapse, every gene on one chromosome is directly opposite the other gene for the same trait. Now, when the paired chromosomes are pulled apart, this attraction may be so strong, that the chromosomes may break, and whole pieces of chromosomes may be interchanged. The process is shown diagrammatically in Fig. 27.

Here is another place where chance can exert its influence, for the point, or points, at which crossing over occurs is determined by chance. Consider three points on the chromosome in Fig. 27. Crossing over will occur more frequently between points A and E than it will between points A and B simply because there is a greater distance over which a break can occur. Therefore it follows that the distance between points on a chromosome is proportional to the frequency of crossing over between them. This idea was first proposed by Sturtevant in 1913. Working on this basis, Morgan and his

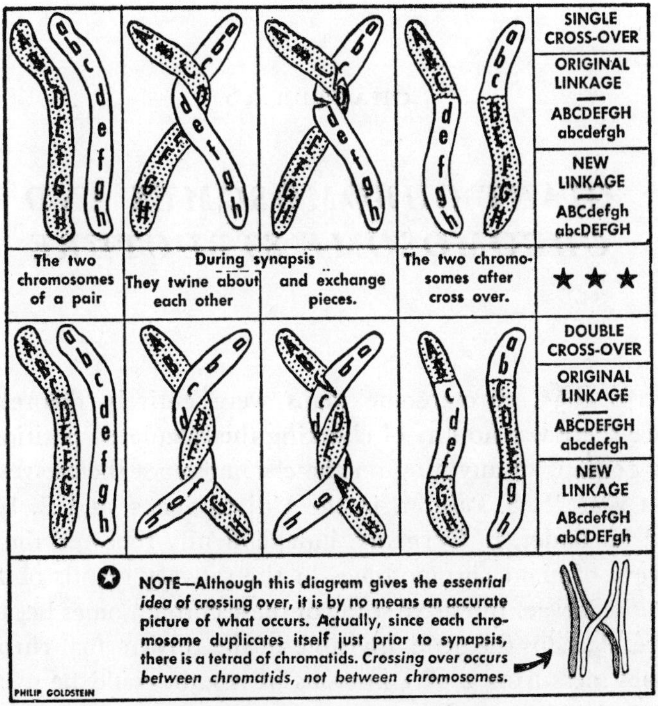

FIG. 27. The effect of crossing over.

co-workers produced maps of the chromosomes in *Drosophila* on which several hundred genes have been located. Subsequently, other experimenters produced similar maps for other organisms such as corn, and some of the smaller plants such as yeast and molds.

CHAPTER 18

GIANT CHROMOSOMES AND CHROMOSOME STRUCTURE

UNTIL 1933, chromosome maps were entirely theoretical, since there was no way of checking the calculated position of the genes with anything on the chromosomes themselves. In that year, T. S. Painter, in the United States, and E. Heitz and H. Bauer, in Germany, independently reported the discovery of giant chromosomes in the salivary glands of *Drosophila* larvae. Intensive study of these chromosomes began at once. Perhaps the gene, invisible in the tiny normal chromosomes measuring 2 or 3 microns in length, could be seen in these giants measuring as much as 400 microns.

It was soon pointed out that the giant chromosomes showed bands or cross striations when stained, and that there was close agreement between the position of these bands on the stained chromosomes and the position of the genes on the maps. T. H. Morgan pointed this out in his Nobel prize lecture at Stockholm on June 4, 1934, when he stated:

> *"Thus, whether or not the bands are the actual genes, the evidence is clear in showing a remarkable agreement between the location of the genes and the location of corresponding bands."*

Whether or not these bands are actually the genes still remains an open question. However, many scientists have attacked it, and perhaps eventually we will know the answer. For example, through the combined efforts of geneticists and cytologists it was shown that in strains of *Drosophila* where certain bands on the chromosomes have been lost, there is a

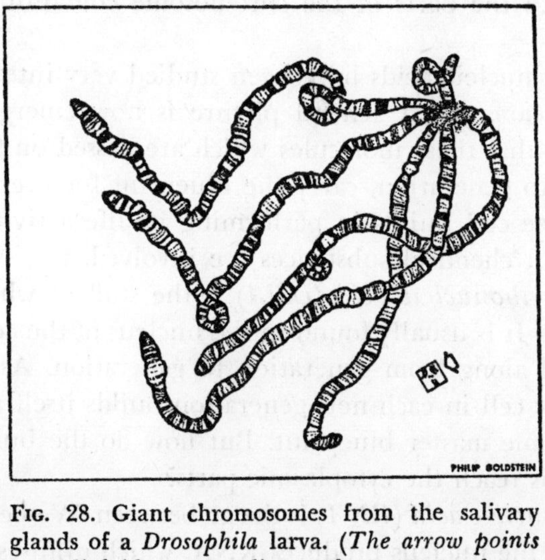

FIG. 28. Giant chromosomes from the salivary glands of a *Drosophila* larva. (*The arrow points to a diploid set of ordinary chromosomes drawn to approximately the same scale.*)

corresponding loss of the specific trait which is associated with that region of the chromosome. Direct chemical analysis shows that chromosomes are rich in nucleoproteins. Feulgen's test, a staining reaction specific for certain nucleoproteins, affects only the transverse bands of the giant chromosomes, indicating that nucleic acid is found only in these bands. This is supported by studies involving X rays and ultraviolet absorption. Protein-digesting enzymes acting on giant chromosomes have dissolved the entire framework, leaving intact only the nucleic acid in the bands. Conversely, other enzymes have digested away the nucleic acid, leaving only the framework of protein.

In general, the work indicates that a chromosome consists of a framework of nucleoproteins, along which nucleic acid is combined at intervals. This accounts for the visible banding of the giant chromosomes. Although the nature of the gene is still unknown, it is believed that gene activity is associated only with the parts of the chromosome containing nucleic acid.

These nucleic acids have been studied very intensively in recent years, and a general picture is now emerging. It is believed that these molecules which are passed on from generation to generation, carry the blueprint for every protein which the cell builds in performing its life activities. Two important chemical substances are involved.

Deoxyribonucleic acid (*DNA*) is the stuff of which genes are built. It is usually found in the nucleus of the cell, and it is passed along from generation to generation. As a result, each new cell in each new generation builds itself according to the same master blueprint. But how do the building instructions reach the cytoplasmic parts?

Ribonucleic acid (*RNA*) is the go-between. We believe that DNA in the nucleus produces RNA, which diffuses out into the cytoplasm. This RNA carries the "instructions" to those parts of the cell which do the building. The RNA molecules may even serve as a model for the formation of the required kinds of proteins.

We have learned a great deal about these nucleic acids, yet we know very little about them. We think that we understand the structure of DNA. We know it to be a complex molecule of giant size as molecules go. But remember; even these giant molecules have never been seen, though they have been studied by chemical means and by X-ray analysis.

In 1953 J. D. Watson and F. H. C. Crick proposed a model of the DNA molecule. Known as the "Watson-Crick Model," this is the best description to date. It pictures the DNA molecule as a double spiral made of smaller molecules linked together in a special way. Think of it as a spiral staircase, with steps made of the different molecules. Since the many small molecules can occur in various sequences, there is hardly a limit to the tremendous number of different combinations which can exist. We believe that the two spirals separate when the nucleus divides. Subsequently each single spiral reduplicates the missing half, and returns to the double spiral condition.

There is experimental evidence that DNA and RNA can duplicate themselves, evidence strong enough to win a Nobel Prize for Ochoa and Kornberg in 1959. Kornberg discovered a bacterial enzyme which could cause a tiny bit of DNA in a mixture of protein molecules to produce more DNA molecules of exactly the same type. Ochoa did the comparable thing with RNA. He used a different bacterial enzyme which he added to a mixture of protein molecules containing a tiny quantity of RNA. More RNA of identical type was synthesized.

Both demonstrations were carried out in vitro (in test tubes) and not in cells. It is assumed that a similar process occurs inside the living cell.

CHAPTER 19

MULTIPLE EFFECTS OF A SINGLE GENE

IN THE PRECEDING sections the gene has been described as a somewhat mysterious something, with the remarkable power of causing the development of some trait in the organism. This idea seemed to carry the implication that each pair of genes affects a single trait. However, this concept is incorrect. In his Nobel prize lecture, Morgan stated:

"As facts accumulated it became evident that each gene produces not a single effect, but in some cases a multitude of effects on characters of the individual. It is true, that in most genetic work, only one of these character effects is selected for study—the one that is most sharply defined and separable from its contrasted character—but in most cases minor differences also are recognizable that are just as much the product of the same gene as is the major effect. In fact, the major difference, selected for classification of the contrasted character-pairs may be of small importance for the welfare of the individual, while some of its concomitant effects may be of vital importance for the economy of the individual, affecting its vitality, its length of life, or its fertility."

To illustrate this situation, it has already been pointed out that *Drosophila* which are pure for vestigial wings live about half as long as normals. Here is a case where we label the gene "vestigial" because of its most obvious effect. Yet this same gene produces other effects, perhaps more important, which weaken one or more parts of the body and so shorten life. There is evidence that in man, too, long life is inherited. It is commonly said that the best way to be sure of living long is to choose parents who lived long, and to stay out of the way of taxicabs. If it is permissible to speculate, perhaps man, too, has genes which shorten life by causing weakening effects on various parts of the body.

Another example deals with the so-called "creeper" fowl, reported by Landauer. Here the "creeper" gene is dominant over normal, but the pure dominant form is lethal. Hybrid creepers are deformed dwarfs with underdeveloped long-bones in their legs. As a result, the birds seem to creep rather than walk. Pure creepers die before hatching from the egg, and show many abnormalities which seem to stem from a defective yolk-sac circulation. Here then, the particular gene named "creeper" obviously affects the leg bones, yet it has a far more important effect on the circulatory system which indirectly causes other traits to be affected.

CHAPTER 20

GENES THAT WORK TOGETHER AND MORE MODIFIED RATIOS

PEOPLE who study genetics are sometimes left with two false ideas:

1. Each pair of genes affects a single trait.
2. Each trait is determined by a single pair of genes.

The first of these ideas, that each pair of genes affects a single trait, was considered on p. 76 and shown to be incorrect. The second idea, that each trait is determined by a single pair of genes, is also incorrect. There is conclusive evidence that many traits are determined by the interaction of two, three, or more pairs of genes, in different pairs of chromosomes. This is known as *factor interaction.* To help understand the examples which appear below, we must remember that when Mendel crossed two dihybrids, he always found among the offspring that

9/16 showed both dominant traits
3/16 showed dominant No. 1 and recessive No. 2
3/16 showed recessive No. 1 and dominant No. 2
1/16 showed both recessive traits

A. Complementary Factors and Reversion to Type

In many organisms there are two or more complementary factors which work together to produce a certain trait. For example, in the sweet pea there are two factors for the formation of red flower color. One of these factors (C) seems to produce a color base, while the second (R) produces an enzyme which activates the color base to produce color. Both C and R must be present together for red color to be produced, since they complement each other. The color base cannot produce color without the enzyme, and the enzyme cannot produce color if it has nothing to work on. Since the C and R genes are carried on different chromosomes, they are inherited independently of each other. But when two sweet peas hybrid for C and R are crossed, instead of getting a normal 9:3:3:1 Mendelian ratio, a modified ratio of 9 red to 7 white results, because only those plants which have inherited both dominants can show red color. The problem is worked out in Fig. 29, by the short method.

Genes	C + R = red C or R alone = white
Problem	CcRr x CcRr ➧ ?
Result	3C < 3R ➧ 9 CR } 9 red 3C < 1r ➧ 3 Cr 1c < 3R ➧ 3 cR } 7 white 1c < 1r ➧ 1 cr

PHILIP GOLDSTEIN

FIG. 29. Complementary factors.

This type of inheritance results in an interesting situation where two strains of white sweet peas that breed true can produce red offspring when crossed (see Fig. 30).

Similar surprising results in the mating of pure strains of other plants and animals provided breeders with a headache as well as a mystery. Lack of understanding was hidden be-

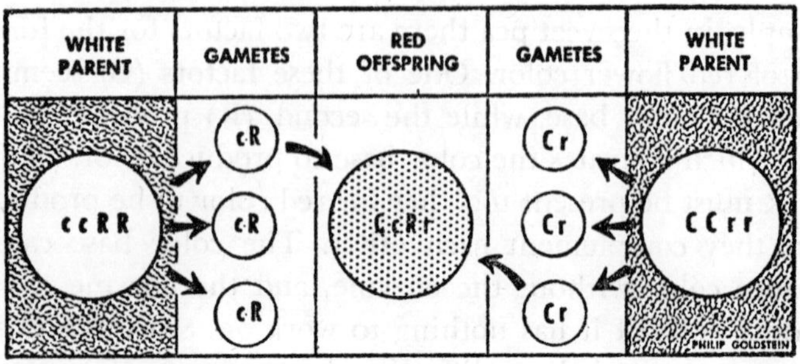

FIG. 30. Reversion in sweet peas. *How two white sweet peas may produce red offspring because of the presence of complementary genes.*

hind the words *throwback,* or *reversion to ancestral type.* The mystery cleared itself up, however, when factor interaction was clearly understood.

Another case which illustrates both complementary factors and reversion to type deals with eye color in *Drosophila.* The wild type of *Drosophila* has red eyes. However, in some inbred strains, different eye colors are noted, as for example scarlet eyes or glass eyes (so called). When these two inbred strains of flies are crossed, all the offspring revert to the wild type having red eyes, showing that the genes for glass and for scarlet complement each other. Evidently, during the process of inbreeding, each of the strains lost one of the complementary factors which is necessary for red eye color, and the missing factor was restored during the cross.

Another interesting case of complementary factors involves the almond. The *bitter* almond contains an organic compound called *amygdalin* and an enzyme called *emulsin.* When bitter almonds are crushed in water, the enzyme causes the amygdalin to break down into glucose (sugar) and benzaldehyde (oil of bitter almond) and HCN (Prussic Acid).

$$\text{AMYGDALIN} \xrightarrow{\text{EMULSIN}} \text{GLUCOSE} + \text{BENZALDEHYDE} + \text{HCN}$$

Although bitter almonds are commercially valuable, they are not edible because the HCN produced is highly poisonous.

By contrast, the sweet almond contains the enzyme emulsin, *but no amygdalin.* Therefore the reaction mentioned above does not occur, and the sweet almond is perfectly edible. It is the nut with which we are all familiar.

Evidently the heredity is the result of the interaction between two complementary gene pairs, the gene for amygdalin production (A), and the gene for emulsin production (E).

The bitter almond that produces both substances must carry both genes in an active form—AAEE, or AAEe, or AaEE, or AaEe. The sweet almond that produces only emulsin must have the active form of the E-gene, but cannot have the active form of the A-gene. Therefore its genetic makeup must be either aaEE or aaEe.

Based on these facts, geneticists predicted that eventually an almond would be found that produces amygdalin, but not emulsin. This prediction has come to pass. Can you determine the genetic make-up which such an almond must have?

Theoretically, there should be still another almond that produces neither emulsin nor amygdalin. What would its genetic make-up have to be?

Probably the same type of inheritance occurs in man, although definite evidence is lacking. If speculation is permitted, perhaps certain types of great musical abilities, or extremes of high intelligence may result from the interaction of complementary genes, each coming from a different parent. This might account for the fact that sometimes outstanding ability appears in one member of a family where other children and parents are not outstanding.

B. Supplementary Factors, or Genes That Modify

Many genes act as modifiers; that is, they alter the effect produced by other genes. For example, in corn there are three complementary genes for color formation, A, C, and R. All three must be present in dominant form in order to produce red color in the aleurone layer of the kernels. In addition, there are several modifying genes (*supplementary genes*) which alter the color formed. Thus, gene Pr changes the red color to purple, provided, of course, that the genes for red color are all present. If A, or C, or R is lacking, then the color is white even though Pr may be present. Another modifier, the color-inhibiting gene I, may prevent the formation of any color, even when A, C, R, and Pr are all present in dominant form.

Coat color in mice is determined by the interaction of at least six pairs of genes. There is a basic color gene (C) without which no color can be produced. The gene A produces the agouti pattern (every hair has a narrow yellow band near the tip) provided C is also present. If A is absent, the mouse will appear solid-colored black or brown, depending on the presence or absence of the factor B. There are other modifying factors for spotting, dilution of color, etc.

C. Duplicate Factors, or Genes That Do the Same Thing

Another type of factor interaction occurs in the presence of *duplicate factors.* In these cases, there are two or more genes in different chromosome pairs, which can do the same thing. Thus, for example, in corn, there is a gene (Y) for yellow endosperm, which is dominant over white. But in another chromosome, there is another unrelated gene (L) which also causes yellow endosperm, and which is also dominant over white. The presence of either or both genes causes yellow endosperm, while in the absence of both, white endosperm is produced. Thus when dihybrids are crossed, the 9:3:3:1 ratio is modified to 15 yellow to 1 white, because only the double recessive plant is white. The problem is worked out in Fig. 31.

In wheat there are three independent pairs of genes for

Genes	L = Y = yellow		
Problem	Ll Yy X Ll Yy ➧ ?		
Result	3L	3Y ➧ 9 LY	15 yellow
		1y ➧ 3 Ly	
	1l	3Y ➧ 3 lY	
		1y ➧ 1 ly	1 white

PHILIP GOLDSTEIN

FIG. 31. Two duplicate factors.

red color. The presence of any one dominant gene produces red color, and the only white plants are those which are pure recessive for all three genes. Here self-fertilization of a trihybrid gives 63 red to 1 white as shown in Fig. 32.

Genes	R = S = T = red			
Problem	Rr Ss Tt X Rr Ss Tt ➧ ?			
Result	3R	3S	3T ➧ 27 RST	63 red
			1t ➧ 9 RSt	
		1s	3T ➧ 9 RsT	
			1t ➧ 3 Rst	
	1r	3S	3T ➧ 9 rST	
			1t ➧ 3 rSt	
		1s	3T ➧ 3 rsT	
			1t ➧ 1 rst	1 white

PHILIP GOLDSTEIN

FIG. 32. Three duplicate factors.

D. Cumulative Factors or Genes That Add Up

Variable traits such as height, weight, intelligence in man; milk production in cows; egg laying in chickens; number of rows of kernels, and length of ears of corn, etc., are often the result of the interaction of many pairs of genes or *multiple factors.* Such traits exhibit wide variability because different individuals differ in their genetic make-up, and in the number of dominant genes for the trait.

For example, in the case of color in wheat kernels, although there are 3 pairs of genes for red, only 1 dominant gene for red is necessary to produce red color. However, the intensity of the redness increases with the number of dominant genes present, so that while a plant having 6 dominants is dark red, a plant with only 5 dominants and 1 recessive is a little lighter, one with 4 dominants and 2 recessives is a little lighter still, and so on, down to the plant which has no dominants and is therefore white. It would seem that a gene for red does something to produce red color. But 2 genes for red do it a little better, 3 a little better still, and so on. The effect adds up with each increase in the number of genes so it is sometimes referred to as *cumulative.*

The inheritance of human skin color seems to follow a pattern similar to this. Let us assume that there are 5 pairs of genes involved, and that their effect is cumulative. Let us assume further, that each dominant gene has a color value of 10, while a recessive gene has a color value of 0.

A "pure Negro" would have 2 dominant genes in each pair (AA BB CC DD EE), and therefore his color value would be 100 or very dark. By contrast, a "pure white" would have only recessive genes (aa bb cc dd ee) with a color value of 0 or very light. Suppose, now, a Negro-white marriage occurred. Each offspring would receive ABCDE from the Negro parent, and abcde from the white parent, giving mulatto (Aa Bb Cc Dd Ee), hybrid for each of the 5 color genes. The color value would be 50, intermediate between the two parents. Intermarriage among mulattoes can produce offspring showing a wide variation in skin colors, ranging from 100, or "pure Negro," through all the intermediate shades (90, 80, 70, 60, 50, 40, 30, 20, 10) down to 0 or "pure white."

Of course we do not know whether 5 pairs of genes are really involved in human skin color. Probably there are many more, some as basic color genes, some as activators, and others as modifiers. To cite only one instance, there seems to be a modifying gene which inhibits all pigment production, regardless of what color genes may be present. This inhibitor may cause a condition known as *albinism* which is characterized by the absence of color in the skin, the hair, and the eyes.

E. Heterosis or Hybrid Vigor

One peculiar effect of multiple-factor inheritance, is illustrated by Mendel's observation that when he crossed tall pea plants averaging 6 feet in height, with short pea plants averaging 1 foot, all the F_1 plants were tall, averaging 7½ feet. The offspring were actually taller than either parent. This condition is known as *hybrid vigor,* or *heterosis,* and it has been noted many times. The mule, hybrid between a jack and a mare, is well known for its great strength (and stubbornness). And all farmers are familiar with the fact that hybrid corn gives far greater yields than inbred pure strains.

There is also evidence of hybrid vigor in human interracial crosses. The famous mutineers of the "Bounty" who settled on Pitcairn Island and intermarried with native Polynesians are a case in point. Studies carried out by H. L. Shapiro show that their children were taller, more vigorous, and more fertile than either parental group. Fischer found the same to be true in South Africa where he studied the offspring of intermarriages between the Boers and native Hottentots. Boas found evidence of hybrid vigor among the half-breed Indians whom he studied.

It is interesting to note that in all these studies the hybrids were found to be intermediate in many other traits such as head shape, skin color, etc. Also, when the hybrids inbred among themselves, there was a gradual loss of the initial hybrid vigor with a regression toward the height, vigor, etc., of the parents.

Perhaps heterosis may be explained on the basis of many pairs of genes for strength, vigor, fertility, etc. One strain has some of these genes in dominant form but lacks others, whereas in the second strain the condition is just reversed. When these two strains are crossed, more genes for size, or vigor, or fertility than either parent possessed individually come together into one zygote. Therefore, the hybrid offspring is larger, more vigorous, more fertile, than either parent. When the hybrid is inbred, various combinations of genes may be produced, thus accounting for the gradual deterioration of the initial hybrid vigor.

CHAPTER 21

MALE OR FEMALE—THE DETERMINATION OF SEX

In days gone by, the birth of a child to a royal family was a very important event. Often a great deal hinged upon whether the child would be a boy or a girl, since in some royal lines only a male could inherit the throne. Obviously, the cause of maleness or femaleness was a pressing question, and ample rewards awaited the physician or philosopher who could assure the king of a male heir. Many theories were offered to explain how sex is determined, and efforts to control sex were based on these theories. Some of them seem quite amusing today, yet they were seriously considered at the time. For example, one theory claimed that the sex of the child was determined by the phase of the moon at the time of conception. Other theories were based on the age of the parents, their relative vigor, origin of the sperm from the left or right testicle, development of the child in the right or left side of the uterus, etc.

Today, though we are not so interested in a male succession, sex determination is still an important consideration. A dairy farmer prefers female calves, since they grow up to be milk producers; a chicken farmer interested in egg production hopes that females will emerge from 99 per cent of the eggs which he sends to the hatchery; and usually a father who already has several daughters would very much like the next child to be a son. Yet, despite the wishes of the dairy farmer and the chicken farmer, approximately equal numbers of males and females continue to be produced; and the poor father is just as likely to be blessed with another daughter as he is to have a son.

The 1:1 sex ratio was meaningless until genetics became a science. Then it was suddenly realized that the ratio was similar to the one which results when a hybrid is crossed with a pure, thus

$$Aa \times aa \rightarrow 50\% \ Aa + 50\% \ aa$$

This led Correns to suggest that sex is an inherited trait, where one type is hybrid and the other pure. Then it was recalled that in 1891 Henking had noted two sperm types in the firefly. One type of sperm had among its chromosomes a very tiny one. The second type of sperm had a large chromosome in its place. In 1902 McClung suggested that this special chromosome pair might have something to do with sex determination, and by 1920 enough evidence had been collected to demonstrate pretty conclusively that this theory was correct.

This evidence led to the *chromosome theory of sex determination* which is summarized in Fig. 33.

Fig. 33 describes a male as having the XY combination of chromosomes, and this theory seemed to be satisfactory for a while. But it was soon noted that in certain insects no Y chromosome existed at all. Although the female was still XX, the male was XO, having only one sex chromosome. Yet he was just as much a male as another kind of organism with the XY combination. This led to the conclusion that the Y chromosome was unimportant; merely so much dead wood. Perhaps only the X chromosomes were related to sex determination. Where 2 X chromosomes existed together, the organism was female, while a single X chromosome (with or without the Y), caused maleness.

1. A special pair of chromosomes called *sex chromosomes* determines sex. (By contrast, all chromosomes other than sex chromosomes are called *autosomes.*)

2. In a female, both sex chromosomes are identical. They are known as *X chromosomes.*	In the male, one sex chromosome is an X, like those in the female; but the second is different, and is called the *Y chromosome.*
3. Thus, the chromosome complement of a female is: AUTOSOMES + XX Human.............. 44 + XX *Drosophila*............ 6 + XX	Whereas the chromosome complement of a male is: AUTOSOMES + XY Human................ 44 + XY *Drosophila*............... 6 + XY
4. Since reduction division occurs during egg production, *every egg contains ONE X CHROMOSOME.*	Since reduction division occurs during sperm production, *two types of sperms are produced in equal numbers:* those which carry ONE X CHROMOSOME, and those which carry ONE Y CHROMOSOME.
5. Therefore, the chromosome complement of an egg is: Human.............. 22A + X *Drosophila*............ 3A + X	Therefore the chromosome complement of a sperm is: Human..... 22A + X or 22A + Y *Drosophila*.... 3A + X or 3A + Y
6. When fertilization occurs *the zygote must receive an X chromosome from the egg,* since every egg carries an X chromosome.	But *the zygote may receive either an X or a Y from the sperm,* depending on which type of sperm fertilized the egg.

7. Since chance determines which sperm fertilizes the egg, if several offspring are produced, it is likely that 50 per cent will be XX or female, and 50 per cent will be XY or male.

X egg + Y sperm → XY → MALE
X egg + X sperm → XX → FEMALE

8. In a few organisms (birds, butterflies, and some fish) the situation is reversed. The male has two similar sex chromosomes (ZZ) whereas the female has two different ones (ZW).

FIG. 33. How sex is determined.

Although the chromosome theory seemed to explain sex determination fairly well, a number of flaws soon became apparent. Around 1922, Bridges discovered a *Drosophila* fly whose cells contained 9 autosomes (3 sets in place of the normal 2 sets) and 2 X chromosomes. Since this fly had 2 X chromosomes, it should be a female if the chromosome theory is correct. But the fly was *not* a female. It was an intersex—neither male nor female. Therefore it was evident that the X chromosomes alone did not control sex, since the presence of an extra set of autosomes upset the balance.

From such evidence, Bridges modified the chromosome theory of sex determination to what he called the theory of *genic balance*. As applied to *Drosophila*, this theory states:

1. There are many genes for maleness, and many genes for femaleness scattered throughout the chromosomes.
2. Every individual has genes for both conditions.
3. The X chromosomes carry a preponderance of genes for femaleness.
4. The autosomes carry a preponderance of genes for maleness.
5. The sex of the organism is determined by the balance between the male and female genes.
6. A normal *Drosophila* fly has 6 autosomes.
7. The excess of male genes on 6 autosomes is greater than the excess of female genes on 1 X chromosome. Therefore 6 autosomes + X(Y) is a male.
8. The excess of male genes on 6 autosomes is smaller than the excess of female genes on 2 X chromosomes. Therefore 6 autosomes + XX is a female.
9. The excess of male genes on 9 autosomes is about equal to the excess of female genes on 2 X chromosomes. Therefore 9 autosomes + XX is an intersex, neither male nor female.

Even this theory does not completely explain all the factors which seem to enter into sex determination. Evidence has been produced that metabolic rate, environmental factors, hormones, etc., all play a part, at least in some organisms. Nor does the theory of genic balance explain how sex reversals can occur. One such case involves a hen which laid eggs for a while. Gradually, she took on male characteristics, and finally fathered several chicks. Autopsy showed that the ovary had been destroyed by tuberculosis. Yet, as far as we know, destruction of the ovary does not alter the chromosome structure of the rest of the cells. If genes were the sole determining factor in sex, this hen could not have become a functioning rooster.

Some fascinating facts about sex determination in man have recently come to light. For example, by examining the cells it is now possible to determine whether a fragment of epithelium or a drop of blood comes from a man or a woman.

This was first noticed in the nerve cells of cats. The nuclei of cells from female cats showed a darkly stained spot which was absent in cells from male cats. The unusual spot was named *the Barr body*. Similar Barr bodies were soon discovered in various cells of other animals, including man; but always in the female of the species. Thus a cheek epithelium cell from a normal woman usually has one Barr body. The comparable cell of a normal man has none.

However, this general rule does not hold in certain kinds of sex abnormalities. Men with *Klinefelter's syndrome* have underdeveloped testes, and tend towards femaleness in many respects. Each of their cells has one Barr body, and a chromosome count of 47 (44 autosomes + XXY) instead of the normal 46. Women with *Turner's syndrome* have tiny ovaries and poorly developed sex traits. Their cells have a chromosome count of 45 (44 autosomes + XO) instead of the normal 46. Certain rare persons have cells with three X-chromosomes; or even four. Their cells contain one Barr body fewer than the number of X-chromosomes.

This table summarizes the relationship between sex chromosomes, outward sex, and Barr bodies.

Sex Chromosomes	*Outward Sex*	*# of Barr Bodies*	*Sex Chromosomes*	*Outward Sex*	*# of Barr Bodies*
XY	MALE	0	XO	FEMALE	0
XXY	MALE	1	XX	FEMALE	1
XXXY	MALE	2	XXX	FEMALE	2
XXXXY	MALE	3	XXXX	FEMALE	3

But there is something more worthy of note. Fig. 33, p. 89 draws an analogy between sex determination in *Drosophila* and man. Obviously this analogy does not hold true all the way. The XO combination of chromosomes in insects (p. 88) produces a male. But in man it produces a female. The XXY combination in *Drosophila* produces a female, as though the Y were totally absent. But in man the Y-chromosome must play a determining part, since a single Y-chromosome causes maleness no matter how many X-chromosomes are present.

Thus, the *Drosophila* type of sex determination (pages 89-90) does not hold in man. The human Y-chromosome seems to push development towards maleness by stimulating the sex glands to develop into testes. In the absence of a Y-chromosome, ovaries begin to develop.

There is also evidence that in a female with two X-chromosomes both are not working simultaneously. It appears that one is "turned on" and the other is "turned off". What "throws the switch" to turn one or the other of the two X-chromosomes on is not clearly understood. However, it is only reasonable to assume that if the two X-chromosomes happen to be carrying opposite alleles for a particular trait, the form of the trait which is visible would be determined by the X-chromosome that is functioning at the time when this particular trait is developing.

CHAPTER 22

THE MALES SHOW IT—SEX LINKAGE

THE PREVIOUS section explained how the X chromosomes are related to sex determination. It must not be assumed from this that X chromosomes transmit sex traits only. On the contrary, genes for many traits which are in no way associated with sex are also carried on the X chromosomes, together with the sex genes. For example, in *Drosophila,* the gene for red or white eyes is on the X chromosome. In man, the gene for red-green color blindness is on the X chromosome. Those genes which are carried by the X chromosome are said to be *sex-linked,* since their inheritance follows certain peculiar patterns tied up with the sex of the organism.

For example, some people are bleeders. They suffer from a hereditary condition known as *hemophilia* in which the blood does not clot normally. But the condition appears *mainly in males.* It is usually considered to be a sex-linked trait, controlled by genes on the X chromosome.

However, a complex process like clotting involves many chemical substances (enzymes from the platelets; calcium; plasma proteins like globulins, thromboplastin, fibrinogen; etc.). If production of each of these factors is gene controlled, then several different genes must be involved. A defect in each separate gene may cause a different clotting defect.

Obviously, *hemophilia* is not a single condition. Perhaps there are as many different kinds of hemophilias as there are factors involved in the clotting process. In 1963, one authority listed eight different bleeding conditions, some inherited as sex-linked traits and others as autosomal traits.

The best known "bleeders disease" is classic hemophilia. It is well known that male bleeders of this type cannot pass the condition on to their sons, but certain females who do not show the defect can do so. Such females are known as *carriers,* since they carry the gene for hemophilia and can pass it on, even though they do not show the defect. Evidently, they also carry a gene for the normal which covers up the defective gene.

This peculiar pattern of inheritance remained a mystery until T. H. Morgan discovered a white-eyed male in a bottle of normal red-eyed *Drosophila* flies whose ancestors had all been red-eyed. Intensive study of the inheritance of this new trait proved that the eye-color gene in *Drosophila* is on the

---

1. Assume that X_n represents a sex chromosome with a gene for normal clotting, and X_h represents a sex chromosome with a gene for hemophilia. Y carries no gene for this trait.

2. A normal female has the genetic make-up X_nX_n	A male bleeder has the genetic make-up X_hY

3. According to the chromosome theory of sex determination, a male child receives an X chromosome from his mother, and a Y chromosome from his father.

4. The normal mother can pass on to her son only X_n	The bleeder father can pass on to his son only Y

5. Therefore every son resulting from this marriage is normal and has the genetic make-up X_nY.
6. A male of this type (X_nY) can never pass on hemophilia even though his father was a bleeder.

FIG. 34. Why a boy cannot inherit hemophilia from his father.

X chromosome. Since a female has 2 X chromosomes, she has 2 eye-color genes. But since a male has only 1 X chromosome, he has only 1 gene for eye color. (The Y chromosome carries very few genes. Most of it is inert material.) As a result, a female shows white eyes only if she has 2 genes for white, whereas a male needs only 1 gene for white to make that trait show.

In general, eye-color inheritance in *Drosophila* follows the same peculiar pattern of inheritance as hemophilia in man. Once the inheritance of eye color in *Drosophila* was clearly understood, much of the mystery about the inheritance of hemophilia cleared up. For example, it became clear why a male bleeder couldn't possibly pass the condition to his sons, while at the same time, every one of his daughters became a carrier of hemophilia (see Figs. 34 and 35).

---

1. The normal mother has the genetic make-up X_nX_n	The hemophiliac father has the genetic make-up X_hY

2. According to the chromosome theory of sex determination, a female child receives an X chromosome from the mother and another X chromosome from the father.

3. The normal mother can pass on to her daughter only X_n	The hemophiliac father can pass on to his daughter only X_h

4. Therefore every daughter resulting from this marriage is a carrier with the genetic make-up

$$X_nX_h$$

FIG. 35. Why the daughter of a bleeder becomes a carrier of hemophilia.

Since normal is dominant over hemophilia, a female with gene make-up X_nX_h is not a bleeder, but she can pass on the trait to half of her sons, even when she marries a normal male. The explanation for this is given in Fig. 36.

1. The carrier mother has the genetic make-up X_nX_h	The normal father has the genetic make-up X_nY
2. According to the chromosome theory of sex determination, a male child receives an X chromosome from the mother and a Y chromosome from the father.	
3. The carrier mother can pass on to her son either X_n or X_h	The normal father can pass on to his son only Y
4. Therefore 50 per cent of the sons will be normal (X_nY) while 50 per cent will be bleeders (X_hY).	

FIG. 36. Why half of the sons of a carrier will be bleeders.

It is interesting to note the *crisscross pattern* of inheritance which is characteristic of sex-linked traits. Hemophilia passes from a man to his daughters to his grandsons, thus skipping from male to female to male. This type of inheritance is known in many organisms, both plant and animal.

Secondary sex characteristics are often classified as *sex-limited traits.* Although the genes for these traits are carried on the autosomes (not on the sex chromosomes), they do not affect males and females in the same way. One example of such a trait is the presence of horns in the males but not in the females of certain species of animals. Perhaps in these cases, gene action is modified by the sex hormones.

A very small group of traits is carried by the Y chromosome. Although the Y is mostly inert, there are a few genes on it. Since Y chromosomes are found only in males, these traits can appear only in males. Such inheritance is usually referred to as *one-sided inheritance.*

In the case of man, the sex chromosomes may be divided

into three parts. There is a portion of the X chromosome which has no counterpart on the Y chromosome. Genes on this part of the X chromosome are sex-linked. Similarly, there is a part of the Y chromosome which has no counterpart on the X. Genes on this part of the Y chromosome are said to be *holandric,* and they are inherited by the male only. Lastly there is a portion of the X chromosome which is homologous with a portion of the Y chromosome. The genes on this section are only partially sex linked. Fig. 37 shows the X and Y

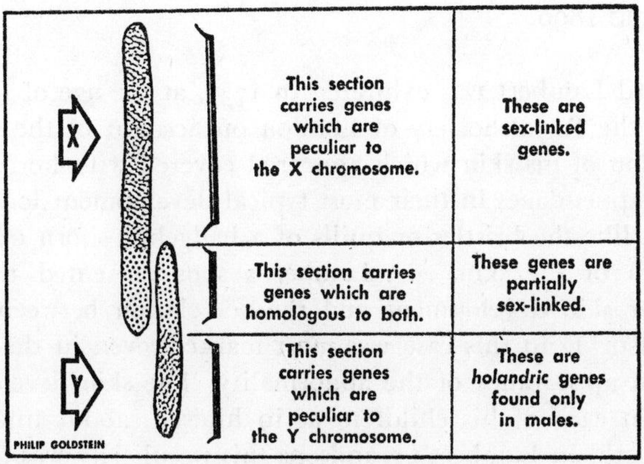

FIG. 37. The sex chromosomes in man.

chromosomes, and the relationship of the three parts mentioned above.

Since only males possess Y chromosomes, only males can inherit the genes carried on the Y chromosome. If such a gene has no counterpart on the X chromosome, it is logical that only males can be affected. This explains the name "holandric" which stems from the Greek (holo = entire + andro = male). Such traits are entirely male in every sense of the word.

Holandric genes are quite rare. At least we do not recognize their expression very often. Only three or four are definitely known in man. One of these produces a condition known as "hairy ears", in which there is a dense growth of long hairs on the ears. Since the trait is controlled by a gene on the Y chromosome, it can only be inherited from the father, and only by the males in the family. In one family it has been traced from father to sons for five or six generations.

Another trait controlled by a holandric gene is known as "porcupine skin". In this trait, the skin over most of the body becomes thick and very rough. Here is a description of the condition and its inheritance written sometime between 1880 and 1890.

"Edward Lambert was exhibited in 1731, at the age of fourteen, before the Royal Society of London on account of the peculiar condition of his skin which appeared covered with horny scales. These appendages in their most typical development looked and rustled like the bristles or quills of a hedgehog shorn off within an inch of the skin. . . . Lambert's sons presented the same peculiar skin development and the correlation between parent and offspring in this case was most marked, even in the date of the first appearance of the abnormality. The skin developed its scales in each of his children, as in himself, about nine weeks after birth. In Lambert's grandsons this peculiarity was also well marked. Two brothers, grandsons of Lambert, were exhibited in Germany on account of their peculiar body covering."

CHAPTER 23

MUTATIONS—GENES ARE SUBJECT TO CHEMICAL CHANGE

ACCORDING TO MENDEL, each hereditary trait is transmitted by a certain physical unit that passes in regular fashion from one generation to the next. Although Mendel did not pinpoint these physical units, later workers placed them in the nucleus, and more specifically in the chromosomes. The term *genes* was applied to these mysterious units, and this name was almost universally accepted and used. A gene was described as a very large and complex organic molecule. Chromosomes were pictured as groups of such molecules held in a definite linear arrangement on a protein framework by some powerful forces that were not clearly understood.

Our knowledge of heredity has advanced rapidly since the days when the gene theory was first proposed. Many modern geneticists are no longer ready to accept the idea that there are discrete physical units such as genes would have to be. We have learned a great deal about the nucleic acids, and their role in heredity (see chapters 37-49). Still, even if genes do not actually exist as such, the gene concept did provide a very valuable base of operations for geneticists. It led to an understanding of the basic manner in which heredity operates. For example, the gene concept made it easy to explain how changes occurred in the heredity of living things.

If a gene is really an organic molecule, it must be subject

to all the laws of chemistry to which other molecules are subject. For example, molecules are capable of undergoing chemical change. What change occurs in a molecule depends upon the conditions under which the change is occurring, and so a molecule may change in different directions at different times. However, if the conditions are kept constant, the same molecular change may occur many times. Often changes which occur are reversible, and since molecules differ in stability, some types will change more readily or more frequently than other types. All of these characteristics of molecules have been shown to hold true for genes, and the following pages will deal with them.

A. Genes Change

Gene changes are known as *mutations.* A mutation can be detected only if it causes a visible change in an organism being studied. Thus, the first mutation to receive intensive study was the change from red to white in the eye color of *Drosophila.* Mutations may occur in the body (*somatic*) cells or in the reproductive cells of an organism. Obviously, a somatic mutation dies with the organism, and so it does not affect heredity. By contrast, a mutation in the reproductive cells may be passed on to the next generation, and the change will be hereditary. Such a hereditary change need not show up immediately, since it may be a recessive trait masked by an unchanged dominant gene. In fact, most mutations are recessive to the wild type. However, even a recessive mutation may eventually appear in some subsequent generation as a result of segregation and recombination (see p. 49).

It is interesting to note that most mutations are harmful to the organism rather than beneficial. Perhaps this is to be expected, since each type of organism has probably evolved the best combination of traits for survival in the struggle for existence. Any change in this combination of traits is more apt to be undesirable than desirable.

B. Multiple Alleles—Genes May Change in Different Ways

Every gene is located at a definite point on a particular chromosome. This very specific position is known as the *locus* of the particular gene. For example, the gene for red eyes in *Drosophila* is located at locus 1.5 on chromosome No. 1. Of course the gene for white eyes is also located at the same locus, since it was formed when the gene for red mutated. Because the genes for red and white are at the same locus, they are known as *allelomorphs* or *alleles* (see p. 18).

However, the gene for red mutated in different directions at various times, because different conditions were called into play. Red also changed to wine, coral, blood, cherry, apricot, eosin, ivory, buff, tinged, ecru, etc. Here, then, is a whole series of genes for eye color, all of which are located at the same locus, and all of which are consequently allelomorphic. Several genes, all at the same locus, are known as *multiple alleles*. This must not be confused with multiple factors, where several genes at different loci affect the same trait (see pp. 78-86).

Since any individual *Drosophila* fly has only two No. 1 chromosomes, a single fly can only have two genes for eye color at 1.5 on chromosome No. 1. But it can have any combination of two genes. It might have two genes for red, or two genes for white, or a gene for red and a gene for white. It might have two genes for eosin, or one for eosin and one for white, or one for eosin and one for red. And the same is true for every other color in the series. All the genes in the series can be arranged in the order of their dominance relative to one another. To illustrate, in the abbreviated list given,

RED
CHERRY
EOSIN
BUFF
WHITE

each gene is dominant over those listed below it, and recessive to those above it. Thus red is dominant over all the others, while white is recessive to all. Cherry is dominant over eosin, buff, and white, but is recessive to red. Buff is dominant over white, but is recessive to the others.

Multiple alleles are found in many organisms, arising as the result of the fact that the gene molecule at a particular locus changed in different directions at various times in the history of the species. Man is no exception. Several series of multiple alleles exist in man. For example, hair form is determined by a series of genes which runs in order of dominance as follows: kinky, curly, wavy, straight. Another interesting series of multiple alleles in man deals with blood type. Here there are three allelic genes, known as A, B, and O. Of these, A and B both are dominant over O, but neither A nor B can dominate the other. Since an individual can have only two of these genes, four blood groups are possible—

Group A—people who have gene combinations AA or AO
Group B—people who have gene combinations BB or BO
Group AB—people who have gene combinations AB
Group O—people who have gene combinations OO

It is these gene combinations which determine what blood group a person belongs to, and what type of blood must be used in transfusion. The fact that these genes are inherited, just like any other genes, is used in carrying out paternity tests (see pp. 135-143).

C. The Same Mutation May Occur Many Times

In a controlled chemical process, many molecules of a certain type may all be made to undergo the same change. Similarly, gene molecules may undergo the same change at different times, if conditions happen to be similar. This leads to the same mutation, and therefore the same mutation may occur many times. An interesting side light on this phenomenon is the fact that the same or similar mutations have been observed in related species of organisms. For example, the Rh blood factor, which is so important in man, was named "Rh" because it was first discovered in the Rhesus monkey. Perhaps such *parallel mutations* occur because of the similar nature of the molecules which make up the genes of related species.

D. Genes May Mutate Back to Their Original Form

Although *reverse mutations* are not frequent, they are known to occur in various organisms. For example, in *Drosophila,* bar-eye mutation reverses itself and mutates back to normal. Carrying through our analogy with chemical changes, this seems to indicate that whatever change occurred in these particular genes was reversible.

E. X Rays and the Rate of Mutation

Genes are extremely stable molecules. Therefore they do not often change. Muller estimated that in *Drosophila,* under normal conditions, the average gene mutates *only once in a million generations.* However, since each gene has a different molecular structure, genes are not equally stable. Some mutate more often or more easily than others. The rate of mutation of several genes in corn was worked out by Stadler. He found that in over one and one-half million gametes tested, the gene called "waxy" never mutated, whereas the color gene R mutated nearly five hundred times per million gametes studied; and the modifying gene for purple color mutated only about ten times per million gametes. Although these figures are only approximations, they suffice to show the wide variation in mutation rate for different genes.

Many attempts have been made to accelerate the rate of mutation by artificial means. Organisms have been subjected to extreme conditions of environment—low temperature, extreme moisture, variations in food, even powerful rays of one sort or another. But evidently the genes are extremely well protected, or they are too stable to be affected, for most of these efforts did not change the rate of mutation. However, around 1927, H. J. Muller announced that he had devised a technique by which the mutation rate in *Drosophila* could be increased to nearly one hundred and fifty times the normal rate. He shot X rays at the reproductive cells, in almost killing doses. Evidently, the X rays penetrated into the cells and caused the genes to mutate by providing the energy necessary for the change. Muller found that many of the mutations which he produced artificially were similar to those which occur naturally. Just as in the case of natural mutations, most of the artificial mutations were recessives, often harmful or lethal to the organism. Reverse mutations occurred, showing that the changes were reversible. Muller was awarded the Nobel Prize in 1946 for his work on mutations.

Almost simultaneously with Muller's announcement of his work on *Drosophila* came the report that L. J. Stadler had used X rays to accelerate the rate of mutations in plants. Later, it was found that radium emanations, neutrons, ultraviolet light, and other forms of irradiation were also effective, as were extremely high temperatures applied at the proper time in the life cycle. In all cases, the closer to lethal the conditions to which organisms were subjected, the more effective these conditions were in producing mutations.

Will the radiations cause changes in the human race?

The atomic bombs dropped in Japan during World War II produced intense radiation. Genetic changes must have occurred in some of the people who survived. How these mutations will affect future generations is still a subject of study. Enough time has not yet elapsed to form a definite judgment. But it is a frightening thought that genetic changes may be an outcome of modern warfare. We *know* that most mutations caused by X-rays are harmful to the species subjected to the irradiation. Is there any reason to believe that mutations in man resulting from atomic explosions will be any less harmful?

CHAPTER 24

OTHER TYPES OF CHROMOSOME CHANGES

IN THE PREVIOUS section, the word *mutation* was used to describe a change in a gene. Originally, the word had a much broader meaning. It was used to refer to any sudden change in the heredity of an organism, no matter how it was brought about. However, many hereditary changes do not involve the actual change in the chemical nature of a gene, and it is customary today to reserve the word mutation for gene changes only. Some of these hereditary changes, other than gene mutations, will now be described very briefly.

A. Deletion and Deficiency; Translocation and Duplication

Sometimes a piece of chromosome breaks off during gamete formation and gets lost. This is known as *deletion,* since certain genes have been removed from the chromosome where they belong. If the deleted piece of chromosome attaches itself to another chromosome, it is known as a *translocation,* since certain genes have been transferred to a location on a chromosome where they do not belong.

In either case, the heredity is altered. Some gametes will be *deficient* in certain genes—those carried on the missing piece of chromosome. Other gametes may have *duplicate* genes—those carried on the extra piece of chromosome which doesn't really belong in the gamete at all. Often, deficient gametes cannot function, or the zygote never develops, perhaps because a gene for some vital trait is lacking. But if no vital traits are involved, a deficient gamete may give rise to an organism which shows some alteration in a visible trait. Or, if an organism has a duplication, one of its traits may be pushed to extreme form.

By careful microscopic examination of the chromosomes of organisms showing such altered traits, or extreme traits, it is sometimes possible to locate the actual point on the chromosome where the deficiency or duplication exists. When this is possible, the gene for the trait in question is definitely proved to be located in the missing or duplicated region of the chromosome, and consequently the location of the gene is known in the normal organism. Thus, deletions and translocations are valuable aids to the geneticist in confirming chromosome maps (see p. 70).

Sometimes two chromosomes which are not members of the same pair exchange pieces. This condition is known as *reciprocal translocation* or *segmental interchange*. This should not be confused with crossing over (see p. 71) which involves the two chromosomes of the *same* pair.

B. Inversion—Pieces of Chromosomes Turned Upside Down

When a piece of chromosome becomes inverted, no genes are gained or lost, but the order of the genes is altered. If the original sequence of genes on a chromosome is

ABCDEFGHIJKLMNOPQRSTUVWXYZ,

and section EFGHIJKLM becomes inverted, the new sequence of genes becomes

ABCD*MLKJIHGFE*NOPQRSTUVWXYZ.

This *inversion* does not directly affect the traits of the organism, since the same genes are still present, but it does cause abnormalities during gamete formation.

Remember that, during reduction division, the two chromosomes of a pair come into close contact (*synapsis*) with each point on one chromosome directly opposite the corresponding point on the other (see p. 34). In the normal case, this can be pictured as follows:

ABCDEFGHIJKLMNOPQRSTUVWXYZ
ABCDEFGHIJKLMNOPQRSTUVWXYZ

Now imagine what happens when a normal chromosome tries to synapse with a chromosome carrying an inverted section. Even contorted positions of the two chromosomes cannot bring about point-for-point correspondence. Yet, the microscope reveals that, during synapsis, chromosomes with inverted sections undergo just such contortions in their efforts to bring about proper matching of genes. Usually, normal gamete formation is interfered with, and certainly, *crossing over can never occur in an inverted part of a chromosome.*

C. Nondisjunction and Heteroploidy—Extra Chromosomes

Occasionally a pair of chromosomes fails to separate during reduction division. This condition, known as *nondisjunction,* gives rise to abnormal gametes, some carrying 1 extra chromosome (n + 1), and some lacking 1 chromosome (n — 1). If a gamete with an extra chromosome unites with a normal gamete, the zygote has 2 full sets of chromosomes, plus 1 extra chromosome (2n + 1). It is now said to be *trisomic* for the particular chromosome of which there are 3. If 2 (n + 1) gametes unite, the zygote is (2n + 2), or *tetrasomic,* for the chromosome of which there are 4. Trisomic and tetrasomic organisms show modifications in traits carried by the extra chromosome, since, for these traits, the number of genes has been increased.

In the early days of genetics, Dr. Albert F. Blakeslee found the plant *Datura stramonium* (Jimson weed) very useful in genetic research because it had a whole series of clear-cut traits that followed the Mendelian pattern. However, he observed that certain *groups* of traits failed to follow the expected pattern. He proved that in these cases there was an unusual number of chromosomes resulting from duplication of a whole set of chromosomes or of a single chromosome to produce a trisomic (2n + 1).

Datura has 12 pairs of chromosomes, and Blakeslee was able to produce 12 different trisomic strains. One strain had 12 pairs of chromosomes plus an extra chromosome #1. Another had 12 pairs plus an extra chromosome #2. Still others had an extra chromosome #3 or #4 or #5, and so forth right down the line. A plant with three chromosomes #7 (trisomy-7) always had the same group of unusual traits. But it differed from a plant with trisomy-5 (for example). Somehow, each different chromosome altered the basic traits of the normal plant in its own characteristic direction.

While Blakeslee's work was interesting, trisomy was considered to be an unusual condition without real significance. But some 35 years later, in 1959, a group of French scientists made the startling discovery that *Mongoloid children have one extra chromosome in their cells.* Mongolism is a congenital defect characterized by upward slanting eyes, short neck, flattened skull, and worst of all, a very low grade of intelligence. According to the French scientists, children with this syndrome (group of traits) are trisomic for chromosome #21.

A human cell normally has 23 pairs of chromosomes (22 pairs of autosomes + one pair of sex chromosomes). For convenience, the autosomes are numbered from 1 to 22 in descending order of size. An extra chromosome #21 consistently appeared in the cells of Mongoloid children. This is referred to as *trisomy-21.*

If trisomy-21 can cause such a dramatic effect, can trisomy of other chromosomes do likewise? We already know of serious defects resulting from extra sex chromosomes (see pp. 90-

92). Other defects have been related to trisomy-17 or 18; and trisomy-13 or 14 or 15. (We are not yet certain exactly which is the extra chromosome in these cases.) There is evidence of partial trisomy, where only a part of an extra chromosome is present; and of translocation trisomy, where the extra chromosome is attached to another of a completely different pair. Can we assume that eventually we may find trisomics for each of the 23 human chromosome pairs, just as Blakeslee found trisomics for each of the 12 pairs of *Datura* chromosomes?

D. Polyploidy—Extra Sets of Chromosomes

It will be recalled that chromosomes come in sets (see p. 29), and that in most plants and animals a normal body cell has 2 sets. However, simple organisms like molds and yeasts may exist as either *monoploids,* with 1 set of chromosomes, or *diploids,* with 2 sets. Diploid yeasts develop when 2 monoploids conjugate, thus providing material for interesting speculation. Is it possible that the first organisms were monoploid, and that diploidy arose with sexual reproduction? Is it possible that a double set of chromosomes is better than a single set? Will 2 genes for each trait enable an organism to survive where 1 will not, insurance, perhaps, in case a vital gene is lost?

However, the story does not end with diploid forms. In many plants, varieties are known where chromosomes run in multiples of some basic number. Thus, there are varieties of wheat having 7, 14, and 21 pairs of chromosomes; and varieties of chrysanthemums having 9, 18, 27, 36, and 45 pairs of chromosomes. It is intriguing to speculate about the origin of the 14-pair wheat by the doubling of the chromosomes in the 7-pair variety; or that the 36-chromosome-pair chrysanthemum is a result of two consecutive doublings of the 9-pair variety.

This idea becomes even more interesting when we consider that chromosome sets have been doubled artificially in plants treated with drugs such as *colchicine* or *camphor*. Normal *cell* division is prevented, although the chromosomes do divide. Therefore diploid gametes are produced, instead of the normal monoploid gametes. If 2 gametes, each 2n, unite, the resulting plant is *tetraploid* (4n), having 4 sets of chromosomes in each cell. By crossing tetraploids with normal diploids, plants with 3n chromosomes (*triploids*) are produced. Using Jimson weed as an example, this gives rise to a series of plants with 12, 18, and 24 pairs of chromosomes. Note the similarity to the wheat and chrysanthemum series.

Reduction division and gamete formation in *polyploid* plants may be abnormal, but triploids show far more abnormality than tetraploids. Usually, tetraploids are stronger, more vigorous, and even larger than normal diploids, so that polyploidy has its commercial values. Tetraploid flowers have been produced for their large size and showiness, and tetraploid yeasts, with extra-large cells, have proved of value as a source of food yeast.

CHAPTER 25

THE FIRST FIFTY YEARS OF GENETICS AND THE GENE THEORY

GENETICS was born in 1900 when Mendel's paper was rediscovered by Correns, Tschermak and DeVries. During the next fifty years (1900-1950) vigorous and productive research involving all kinds of animals and plants brought the infant science to maturity.

The stress was on the mechanics of heredity. Geneticists of that era wanted to know what mechanism enabled traits to be transferred with such regularity and precision. As information piled up and the mechanical process became clearer, they found they could predict the results of various matings, and improve domestic plants and animals almost at will.

This knowledge and understanding did not result from the achievements of any one individual; or of any single group of individuals. Rather, it was the accumulation of bits of information contributed by scientists from every corner of the civilized world, which gradually grew and took form, until finally it crystallized into the gene theory of heredity. Still, the names of Mendel, Morgan, and Muller stand out. Mendel, Morgan, and Muller—the three M's of heredity, the giants of the first 50 years.

First came Mendel, the founding father of the whole science.

He laid down the rules. He established the fundamental method of attacking problems of heredity, and the basic technique of analyzing the results. Essentially, Mendel's method consists of the steps shown in Fig. 38.

1. Locate two organisms of the same species which differ in a specific trait.
2. Mate these two organisms to produce an F_1 generation.
3. Inbreed the F_1 generation.
4. Classify the F_2 organisms with regard to the trait being studied.
5. Analyze the results statistically.
6. Once a trait has been studied intensively enough, the information obtained can be used to predict the results of subsequent matings involving this trait.

FIG. 38. Mendel's Method.

The years which followed were years of testing—a period when Mendel's classical method was tried on various traits in all sorts of plants and animals. Morgan was the spark plug for a whole army of investigators who applied this method to the study of heredity in *Drosophila.* He used Mendel's method, but with improvements and refinements which would have made Mendel's eyes open wide in astonishment. Muller was one of Morgan's students who went on to disinguish himself independently with his work on mutations.

It was not long before the intensive work in heredity was being organized and systematized. And soon, the Morgan group succeeded in putting the known facts together into a theory which attempts to explain the mechanism by which heredity operates. This is known as the Gene Theory of Heredity. It is summarized in Fig. 39.

1. Hereditary traits are determined by tiny particles called genes.
2. These particles are transmitted from one generation to the next during the process of reproduction.
3. The genes are strung in single file along the length of nuclear structures called chromosomes.
4. Each gene has a definite position, or locus, on the chromosome in which it resides.
5. The gene at a given locus on a chromosome may take one of several different (allelic) forms.
6. Each gamete has a complete set of genes carried in a single set of chromosomes (monoploid).
7. When two gametes unite, the resulting zygote receives two complete sets of chromosomes carrying two complete sets of genes (diploid).
8. When the zygote develops into an organism, each cell receives two complete sets of chromosomes carrying two complete sets of genes.
9. The interaction of the two sets of genes contained in each cell determines the characters which show in the organism.
10. When this new organism produces gametes, reduction division causes the paired chromosomes to separate, so that only one member of each pair goes to any one gamete.
11. This separation of chromosomes provides the mechanism by which allelic genes are separated from one another.
12. Since each gene retains its own identity at all times, allelic genes separate out in pure form.
13. Chance determines which gametes unite to produce the next generation. Therefore it is chance which determines the recombination of the segregated genes.
14. The chance recombination of genes results in the various Mendelian ratios.

FIG. 39. The mechanism of heredity—the gene theory.

The gene theory suggests that traits are inherited with such regularity because their genes are transferred according to a very mechanical pattern. Sudden appearance of strange and unexpected traits (mutations) is assumed to be the result of sudden changes in specific genes. But what causes the genes to change?

Towards the end of the 1900-1950 era, Muller and others proved that exposure to X-rays and/or other high energy radiation greatly accelerates the mutation rate. Evidently radiation is harmful to genes. It damages them, or inactivates them, or alters them in some way. If the organism survives the change, the altered gene may show itself as a new trait.

* * *

The gene theory describes the method by which genes are passed from one generation to the next. But what do the genes do? Mendel assumed that each hereditary determiner (he called them factors) produced one character in the organism. He believed there was a one-to-one relationship between the determiner and the trait. Modern geneticists do not accept this theory. The evidence seems to show that many genes interact to produce a given character, and that many different characters of the organism are affected by a single gene.

But how do the genes produce the characters? This is another question. Here the classical geneticists cannot produce an answer, because Mendel's method cannot possibly provide one. A new approach to the study of heredity must be devised, built upon the groundwork provided by Mendelian genetics, but with new methods of attack on the problem. Such new approaches to heredity are described in chapter 26.

CHAPTER 26

BIOCHEMICAL GENETICS—THE BIOCHEMICAL CONCEPT OF THE GENE

After studying the preceding pages, the reader may well be impressed by the amount of information about heredity which has been accumulated. The work of Mendel, Morgan, and hosts of others has explained the mechanical aspects of heredity so clearly that geneticists can predict the results of matings with great accuracy and can apply their genetic methods to the improvement of plants and animals.

Geneticists speak glibly of genes doing this, and causing that. But there is an annoying fly in the ointment. Every once in a while, some skeptic asks, "*What is a gene?*" A simple question indeed, but how embarrassing. Despite all the accumulated knowledge about gene behavior, nobody can answer a question as fundamental as "*What is a gene?*"

"We don't know," the geneticists are forced to admit. "But we can tell you how genes behave."

"Perhaps you can," reply the skeptics. "But can you tell us how genes bring about whatever effect you say they cause?"

Again the geneticists are on the defensive, because unfortunately, though they can speak of a gene for brown eyes, they are unable to explain how or why this gene causes brown eye color to develop in the body. Yes, geneticists are on solid ground when they deal with the mechanical aspects of inheritance. Their position here is secure; built up and supported by a great deal of experimental evidence. But they are skating on thin ice when they attempt to explain how genes affect the structure and functions of living things.

The gene is usually described as the unit of heredity, although its nature is not clearly understood. In some unknown manner, the gene causes the differences between the alternate ways in which a trait may appear. The basic method for studying genes was established by Mendel. This method begins with the location of traits which appear in two contrasting forms. Organisms showing the contrasting forms of a trait are mated, and their offspring are inbred so that segregation and recombination of the allelic genes can be determined.

With this method, a specific gene can be identified only if it exists in two or more contrasting (allelic) forms. The reason for this is fairly obvious. Let us assume for a moment that all *Drosophila* flies showed red eyes. The location of this eye-color gene could never be determined, because there is no basis for study—there are no crosses to make involving eye color. The geneticist must wait for a mutation to occur, such as the appearance of a white-eyed fly. Once this happens, he can cross red- and white-eyed flies, and follow up with the rest of the procedure. Of course, the classical geneticist can assist nature a little. He can speed up the rate of mutation and he can bring about other chromosome changes by using X rays, mustard gas, neutrons, colchicine, or other agents. But he still cannot study a gene which exists in one form only.

Yet, the biochemical geneticists insist that a gene which exists in one form only is just as worthy of study as a gene which exists in several forms. As a matter of fact, the most fundamental genes are the ones which exist in single form. For example, it is reasonable to assume that all vertebrate animals have genes which cause the backbone to develop.

Yet these genes can never be studied by Mendel's method, since they do not exist in allelic forms.

Furthermore, organisms which reproduce asexually must have genes for their traits, but reproduction without mating does not allow for the crosses required in the classical method of studying heredity.

For these reasons, they conclude that the large majority of genes remain undiscovered, and will continue to remain unknown until chemical methods are used to expose them. The biochemical attack aims to define such genes in chemical terms, and to relate them to the actual chemical processes which they control in the body.

The results of such biochemical attacks have been very gratifying. They seem to indicate that each gene is very specific, controlling the one chemical reaction in the body for which it is specifically fitted. This it does by producing an appropriate enzyme to make the reaction go. But in addition, each gene also duplicates *itself* during nuclear division.

Therefore, the gene really has a dual function. First, it makes exact copies of itself to send forward to future generations. Second, it produces another chemical substance which participates in some specific body process.

Although Archibald Garrod first conceived the idea of genetic control of body metabolism, it remained for George Beadle and Edward Tatum to develop it into a science. Today, biochemical genetics is an effective weapon in the study of heredity. A few examples have already been described (chap. 10). The next section will cite some others.

CHAPTER 27

SPECIFIC SYNTHESIS—HOW GENES ACT

A. Alkaptonuria—a Hereditary Defect

The case of the urine which turns black has already been described (chap 10). Garrod showed this hereditary condition to be the result of the inability of the individual to destroy a substance called alkapton which is in his urine, and which blackens upon exposure to air. Normal people produce

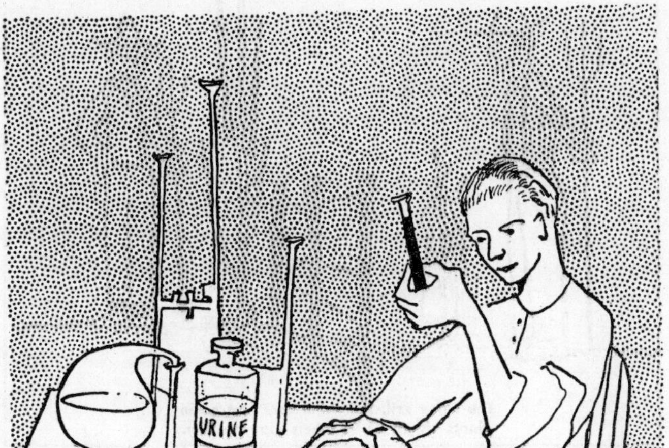

The urine turned black!

an enzyme which decomposes alkapton, and therefore normal urine does not blacken. Since the ability to produce this enzyme is inherited, it would appear that the gene must be the agent which causes the enzyme to be produced. Or perhaps the enzyme is a copy of the gene molecule produced by the gene and released into the blood stream. In any event, *a hereditary defect has been traced to defective enzyme synthesis, associated perhaps with a defective gene.*

B. Susceptibility to Disease

The importance to the farmer of disease-resistant varieties of crop plants is well known. Therefore, geneticists have tried to develop strains of plants which show hereditary resistance,

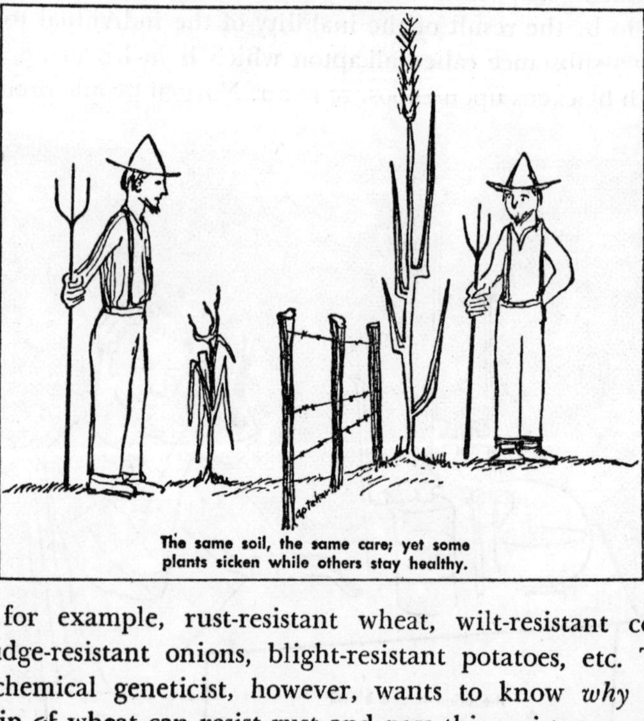

The same soil, the same care; yet some plants sicken while others stay healthy.

as, for example, rust-resistant wheat, wilt-resistant corn, smudge-resistant onions, blight-resistant potatoes, etc. The biochemical geneticist, however, wants to know *why* one strain of wheat can resist rust and pass this resistance on to its offspring, whereas another strain cannot. In a few cases, the explanation has been worked out.

For example, there are some varieties of onions with white scales, and others with colored scales (red or yellow). Colored is dominant over white and is inherited through the action of a single pair of genes. However, onions with white scales are subject to onion smudge disease, whereas colored onions are immune, even when the spores of the fungus are inoculated into the onion. The search for an explanation led to the discovery that the scales of colored onions contain two substances (catechol and protocatechuic acid) which effectively prevent the germination of the spores of smudge disease. The scales of white onions lack these substances. Evidently the gene for colored scales also causes the production of the fungus-inhibiting substances. This is another example where *the gene is involved in a specific chemical synthesis.*

A similar case dealing with hereditary wilt resistance in tomatoes has been reported recently. It was found that wilt-resistant varieties produce a substance called "tomatin," which inhibits the growth of the fungus in laboratory tests. Here also, resistance to a disease is produced by *the gene-controlled synthesis of a specific chemical substance.*

C. Pigment Production

The production of black color in rabbits has already been explained as resulting from a series of chemical changes in which an enzyme is involved (p. 45). This series of changes may be summarized by the following equation:

$$\underset{\text{(an amino acid)}}{\text{TYROSINE}} \xrightarrow[\text{(an enzyme)}]{\text{TYROSINASE}} \underset{\text{(a pigment)}}{\text{MELANIN}}$$

The determining factor in melanin production seems to be the ability of a specific gene to synthesize tyrosinase.

In the guinea pig the situation is quite similar. According to Sewell Wright, there are two primary pigments, a melanic (black) pigment, and a xanthic (yellow) pigment. Both seem to be developed from the same basic substance, perhaps tyrosine, by the action of different enzymes, each produced by a different gene. In addition, several other genes are involved which modify the result by suppressing the action of either or both enzymes, or by localizing it to certain areas of the skin or hair.

Production of eye pigments in *Drosophila* has been studied very intensively by Sturtevant, Ephrussi, Beadle, and others. There appear to be two separate pigments, red and brown, which are inherited independently. Each is produced by a separate chain of chemical reactions, with a common step somewhere along the line. According to biochemical geneticists, each step in the chain is controlled by a specific enzyme, produced by a separate gene. If the enzyme which controls the common step is lacking, no pigment is produced and the eyes are white. Thus, the original white-eyed mutation arose when a certain gene lost its ability to produce its specific enzyme. If any of the other enzymes in the chain are lacking, other modifications in eye color are produced.

Similar chains of reactions have also been worked out for the production of pigments in plants. Here also, each step in the chain is brought about by the action of a specific enzyme. Each enzyme is produced by a specific gene.

In all the cases described, *the gene causes its effect by releasing the specific enzyme needed to stimulate or initiate a certain chemical process in the body.*

D. Human Blood Groups

Inheritance of the human blood groups has already been described (p. 102). It involves one pair of genes, but the gene may appear in three allelic forms—A, B, and O. Of these alleles, gene A produces a chemical substance called antigen A, and gene B produces a similar substance called antigen B; but gene O is inactive and produces no antigen. Since these antigens are found in the red corpuscles, human bloods may be divided into four groups according to the antigens present. This is shown in Fig. 40.

Antigens produced	*Blood group*	*Genetic make-up of individuals in this blood group*
A only	A	AA or AO
B only	B	BB or BO
A and B	AB	AB
None	O	OO

FIG. 40. The blood groups of the *ABO* series.

It has been demonstrated recently that there are really two separate A antigens: A_1 and A_2; so that actually there must be four allelic genes (A_1, A_2, B, O) and six blood groups (A_1, A_2, B, A_1B, A_2B, and O).

In addition to this A-B-O series, other blood groups are also known: the M-N series, the P series, and the Rh series. Each series is inherited independently of the others.

The M-N series involves one pair of genes with two allelic forms, each of which produces its own antigen (see Fig. 41).

Antigens produced	*Blood group*	*Genetic make-up of individuals in this blood group*
M only	M	MM
M and N	MN	MN
N only	N	NN

FIG. 41. The blood groups of the *MN* series.

The P series also involves one pair of genes, but in this case one allele (P) produces an antigen, whereas the other (p) produces no antigen as shown in Fig. 42.

Antigens produced	*Blood group*	*Genetic make-up of individuals in this blood group*
P	P	PP or Pp
None	p	pp

FIG. 42. The blood groups of the *P* series.

The Rh series involves a more complicated system of genes. There are two separate theories which attempt to explain the mode of inheritance of these genes. However, both theories agree that there are three important Rh antigens involved, each with its comparable Hr antigen, and that the presence or absence of these antigens determines whether a person is Rh positive or Rh negative. In general, about 15 per cent of the white population is Rh negative, and 85 per cent is Rh positive.

Dr. Alexander S. Weiner is one of the outstanding American workers in the field of blood grouping. He believes that there is a series of eight multiple alleles involved in the Rh series. He designates these eight genes r, r′, r″, r^y, R^0, R^1, R^2, and R^z. Since these are allelic genes, it means that each one occupies the same position, or locus, on the chromosome which carries the Rh trait. Therefore, any given individual carries two genes for Rh factor—one which he received from his father, and one which came from his mother. The Rh group to which the individual belongs is determined by the particular pair of genes which he carries. In all, there are 36 possible combinations of Rh genes, and they fall into Rh-positive or Rh-negative categories as indicated in Fig. 43.

The second theory is based on the work of British scientists R. A. Fisher and R. R. Race. The Fisher-Race theory assumes that there are three pairs of genes involved, designated as Cc, Dd, Ee. One gene of each pair (C, D, E) produces the Rh antigen, while its allele (c, d, e) produces the comparable Hr antigen. This theory is complicated by the fact

Antigens produced	*Blood group*	*Genetic make-up of individuals in this blood group*
Rh_0	Rh+	Any one of 26 combinations such as R^0R^0 or R^0r
rh′ or rh″	Rh−	Any one of 10 combinations such as rr or r′r

FIG. 43. Positive and negative bloods in the *Rh* series.

that all three genes are linked on the same chromosome, so that any single chromosome may carry one of the 8 gene combinations, indicated in Fig. 44.

Since an individual has two such chromosomes in his cells, there are actually 36 genetic types with regard to Rh and Hr antigens. Of these, people who are genetically cde:cde are

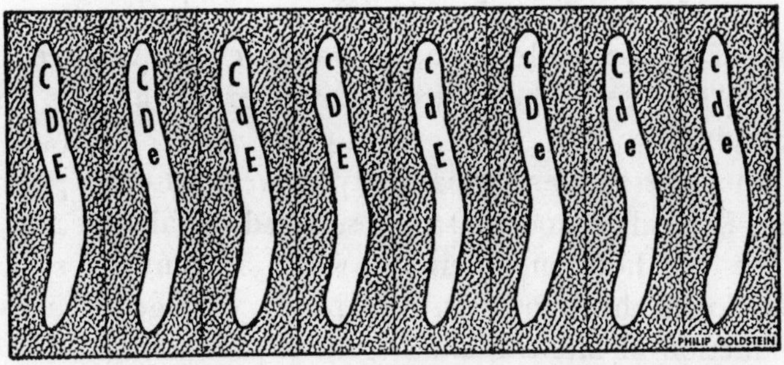

FIG. 44. The eight gene combinations possible in the Fisher-Race theory to explain the inheritance of the *Rh* factor.

completely Rh negative. But since the Rh antigens are not equally strong (Rh_0 produced by the D gene is the only strong one), other genetic types are also considered to be Rh negative for practical purposes (see Fig. 43).

It is noteworthy that all these blood groups result from the fact that *certain genes can cause the production of specific substances (antigens).* It has even been speculated that perhaps the antigen is a "carbon copy" of the gene which produced it.

E. Conjugation in Protozoa

The work of Sonneborn and others has proved beyond doubt that there are hereditary mating groups in *Paramecium* and in other protozoa just as there are plus and minus strains in molds. A plus strain of mold will conjugate with a minus strain but not with another plus. Nor will two minus strains conjugate. The biochemical geneticist is not satisfied with this. He wants to know what chemical difference exists between the opposite strains, and how heredity brings this difference about.

Kimball worked this out for a protozoan called *Euplotes.* He found that the "sex" reaction of *Euplotes* is determined by one pair of genes. But this gene for "sex" can appear in three allelic forms, so that six different mating types are possible. If we use the letters a, b, c to represent the three forms of the "sex" gene, the different mating groups are: aa, bb, cc, ab, ac, bc.

When these different mating types were raised in pure culture, Kimball found that they secreted certain specific substances (sex hormones, perhaps) which accounted for the sex behavior of the organisms. Apparently, each gene causes the production of one substance; gene a produces substance A, gene b produces substance B, and gene c produces substance C. As a result,

an organism having two like genes produces only one substance, thus	whereas an organism having two different genes produces two substances, thus
aa → A	ab → A and B
bb → B	ac → A and C
cc → C	bc → B and C

When two pure cultures are mixed, conjugation never occurs between individuals that produce the same substance, whereas individuals producing different substances do conjugate. Fig. 45 explains which mating groups will conjugate, and which will not.

An individual of mating group	*Will conjugate with another of group*	*But not with one of group*
aa	bb, bc, cc	aa, ab, ac
bb	aa, ac, cc	bb, bc, ab
cc	aa, ab, bb	cc, ac, bc
ab	cc only	aa, bb, ab, ac, bc
ac	bb only	aa, cc, ab, ac, bc
bc	aa only	bb, cc, ab, ac, bc

FIG. 45. The mating groups in *Euplotes.*

Here then, *the gene is responsible for the production of a specific substance* which is directly involved in the mating behavior of the organisms.

F. The Synthesis of Vitamins and Amino Acids

It is a well-known fact that rats do not ordinarily suffer from scurvy, even when their diet provides no vitamin C. This is explained on the basis that rats can synthesize all the vitamin C they require. By contrast, other mammals, including man, must be provided with dietary vitamin C if they are to avoid serious illness. It is also known that certain amino acids, such as phenylalanine, must be provided in the diet, since they cannot be synthesized in the body, whereas some amino acids can be manufactured in the body from other nutritional materials. For example, if the body has phenylalanine, it can make tyrosine.

The biochemical geneticist offers a relatively simple explanation for this situation. According to him, there is a gene for each specific synthesis. Rats have a gene which in some way causes the chemical reactions necessary to produce vitamin C. Other mammals lack this gene, either by loss, by change, or because they never developed it. The same is true of the amino acids. Where there is a specific gene, synthesis is carried on, but where there is no gene, the synthesis does not occur, and this amino acid becomes essential in the sense that it must be supplied by the diet.

In order to test this hypothesis of a specific gene for a specific synthesis, G. W. Beadle and E. L. Tatum carried out an interesting series of experiments. They chose the red bread mold (*Neurospora*) as their experimental material because it is easy to grow in the laboratory; it can synthesize all the B vitamins (except biotin) and many of the amino acids; its life cycle lends itself to the type of study to which it was to be subjected. The life cycle is described briefly below.

1. The vegetative mold is haploid (7 chromosomes).
2. It reproduces asexually by haploid spores (7 chromosomes).
3. If molds of two opposite sexes are grown together, conjugation occurs, giving a diploid zygote (14 chromosomes).
4. The zygote develops into an ascus (spore case) with 8 haploid (7 chromosomes) spores arranged in single file. (Reduction division occurs during the formation of these spores.)
5. Each spore can germinate into a haploid vegetative mold.

The method which Beadle and Tatum used to test the hypothesis is outlined in simplified form in Fig. 46.

Using this method on about sixty thousand single-spore cultures, Beadle and Tatum produced at least forty different strains of *Neurospora, each deficient in one specific synthetic ability, and each passing the deficiency to its offspring*. Spe-

1. *Neurospora* grows well on a *minimal medium* consisting of agar, inorganic salts, biotin, and a disaccharide or fat.

2. The asexual spores (conidia) are subjected to X RAYS, ULTRAVIOLET, or NEUTRONS.

3. The irradiated conidia are added to a normal culture of the opposite sex. Fusion occurs, and sexual spores are produced.

4. Single spores are selected from the ascus and grown on a *complete medium* consisting of agar, inorganic salts, malt extract, yeast extract, and glucose. This medium provides all the essential growth factors, so that the spores germinate and grow well, and asexual reproduction occurs.

5. Single asexual spores are now transplanted to separate culture dishes containing the *minimal medium* (see step 1). There are two possible results:

a. If the mold grows well, it indicates that no genes have been altered, since the mold can still synthesize all the vitamins and amino acids. Such molds are eliminated from further consideration.

b. If the mold fails to grow, it indicates that the ability to synthesize some essential factor has been lost. Such molds are saved for further study.

6. Now single growth factors are added to samples of the mold which no longer can grow on the *minimal medium.* If the mold has lost the ability to synthesize a single factor, it should grow normally when the correct factor is added to the minimal medium.

FIG. 46. Summary of the method used by Beadle and Tatum in their work with *Neurospora.*

cific strains were established which had lost the ability to synthesize the following:

B vitamins	*Amino acids*
Thiamine	Arginine
Riboflavin	Isoleucine
Niacin	Leucine
Pyridoxine	Lysine
p-Amino benzoic acid	Methionine
Inositol	Phenylalanine
Choline	Proline
Pantothenic acid	Threonine
	Tryptophan
	Valine

Here is conclusive evidence that each substance is synthesized under the control of a specific gene, *for mutation of one gene in no way altered the ability to synthesize the other substances.*

G. Sickle-Cell Anemia

In 1910, Dr. James B. Herrick discovered a new disease. It is called SICKLEMIA or SICKLE-CELL ANEMIA because the red corpuscles twist into crescent-shaped sickles as you can see in Fig. 47. The distorted corpuscles twist out of shape be-

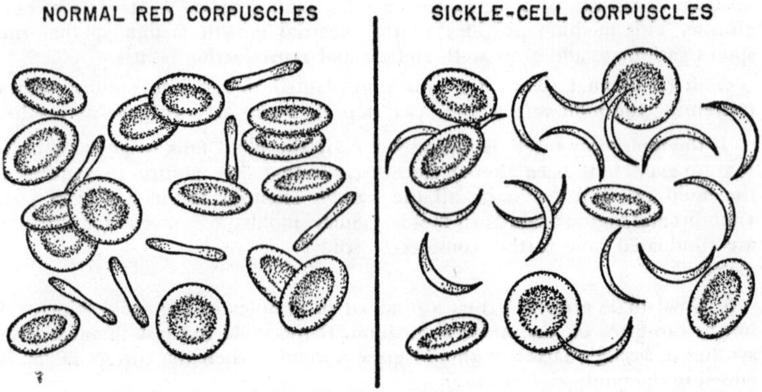

FIG. 47. Comparison of normal red corpuscles with sickle-cell corpuscles.

cause they contain an unusual kind of hemoglobin called Hemoglobin-S in contrast to the normal Hemoglobin-A.

The mode of inheritance of sickle-cell anemia was worked out in 1949 by Dr. James V. Neel. He found that there were two allelic forms of the gene for hemoglobin production.

Gene A causes production of normal hemoglobin-A.
Gene S causes production of sickle-cell hemoglobin-S.

Therefore there are three possible genotypes or genetic make-ups.

AA—if the normal gene is inherited from both parents.
AS —if the normal gene is inherited from one parent, and the sickle-cell gene inherited from the other.
SS —if the sickle-cell gene is inherited from both parents.

How these three genotypes are produced is illustrated in Fig. 48.

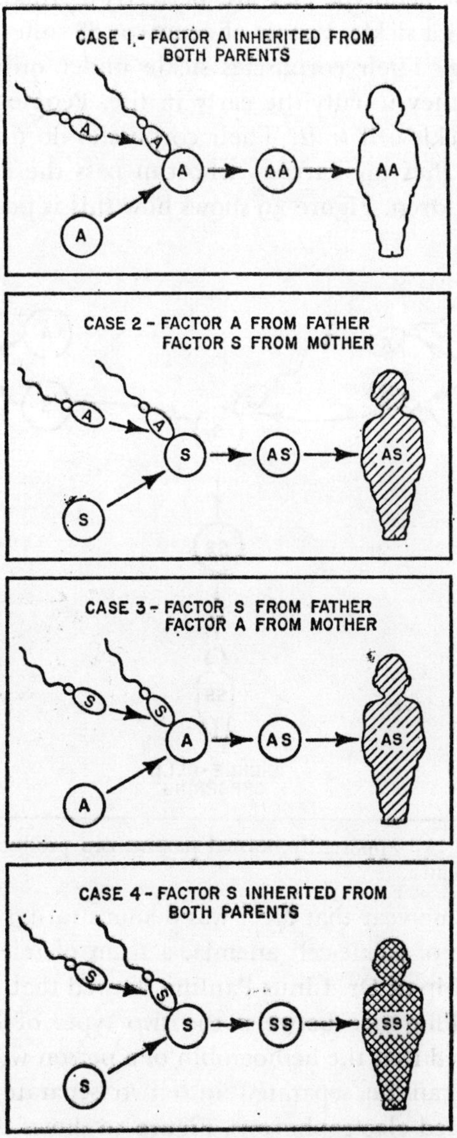

FIG. 48. Inheritance of the sickle-cell gene.

People of genotype AA are perfectly normal. Their corpuscles do not sickle. People of genotype SS suffer from sickle-cell *anemia.* Their corpuscles sickle under ordinary conditions, and they usually die early in life. People of genotype AS carry sickle-cell *trait.* Their corpuscles do not ordinarily sickle, but they are carriers who can pass the condition on to their children. Figure 49 shows how this is possible.

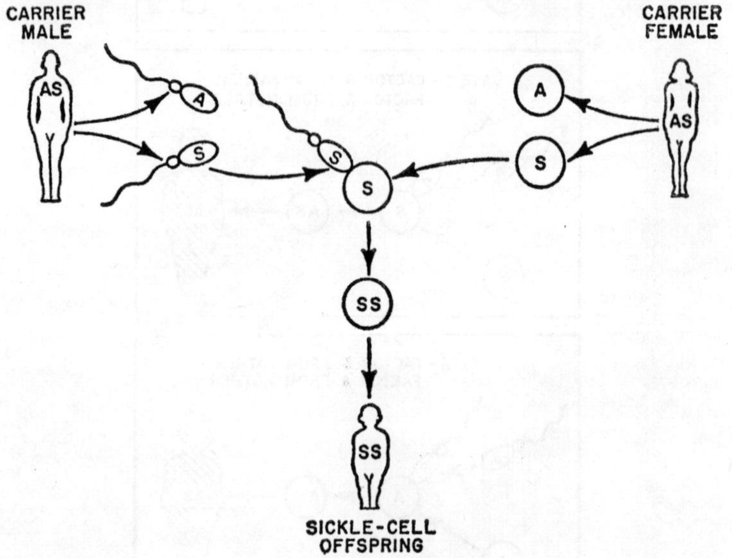

FIG. 49. How two apparently normal parents can produce a child with sickle-cell anemia.

In the same year that Neel was demonstrating the mode of inheritance of sickle-cell anemia, a team of scientists under the leadership of Dr. Linus Pauling showed that there was an electrical difference between the two types of hemoglobin. They showed that the hemoglobin of a person with sickle-cell trait (AS) can be separated into two separate types by a method called electrophoresis. Figure 50 shows that one type is identical with hemoglobin-A and the other type with hemoglobin-S.

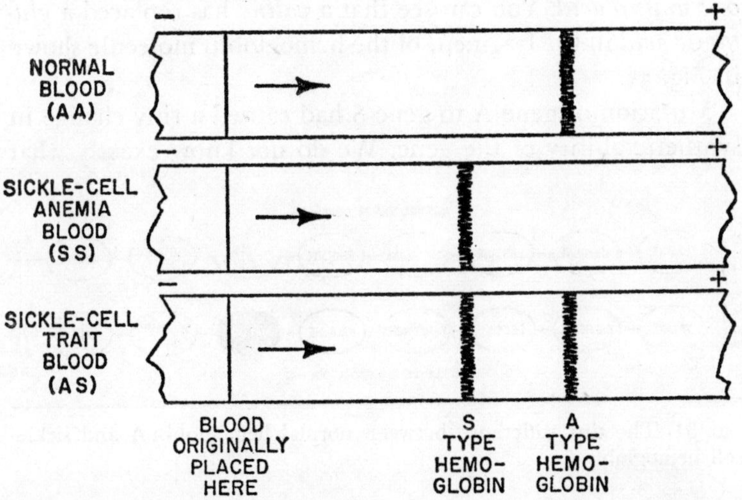

FIG. 50. When subjected to electrophoresis, sickle-cell blood separates into two bands.

This is to be expected if each gene synthesizes its own specific product. The normal gene produces hemoglobin-A, while the mutated gene synthesizes hemoglobin-S. Thus, a person of genotype AS produces a mixture of both types.

The actual chemical difference between the two types of hemoglobins was finally worked out between 1957 and 1961 by Dr. Vernon M. Ingram. He applied a brand new method of analysis that he had invented. It was called *"molecular fingerprinting."* Dr. Ingram showed that in the chain of about 300 amino acids making up the protein part of a hemoglobin molecule, *the two hemoglobins differed in exactly one amino acid.* You can see that a *valine* has replaced a *glutamic acid* in the fragment of the hemoglobin molecule shown in Fig. 51.

Mutation of gene A to gene S had caused a tiny change in synthetic ability of the gene. We do not know exactly what

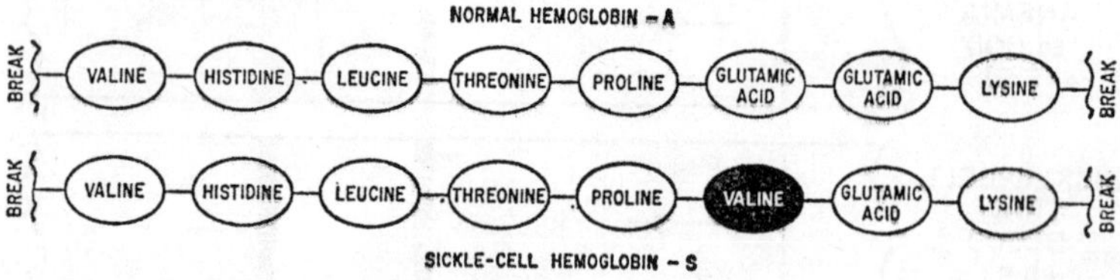

FIG. 51. The tiny difference between normal hemoglobin-A and sickle-cell hemoglobin-S.

change occurred in the gene when it mutated, or how the change was translated into an altered protein. We know only that the single amino acid that was altered makes a tremendous difference in the ability of the hemoglobin to function properly.

A more complete account of the sickle-cell story may be found in pages 34 to 67 of the book *Triumphs of Biology,* by Philip Goldstein, Doubleday and Co., New York, 1965. This includes original accounts of the discovery of sickle-cell anemia and how its heredity was traced and analyzed.

CHAPTER 28

WHO IS THE FATHER? CAN BLOOD TELL?

BEFORE 1900, a blood transfusion was a dangerous operation. It was just as apt to kill the patient as to help him, because when certain bloods are mixed, the serum of one blood will make the corpuscles of the other blood clump together. Such a clump of cells floating around in the blood stream may have disastrous effects if it happens to clog up an important blood vessel somewhere in the body.

In 1900, Karl Landsteiner demonstrated that human bloods fell into four classes based on the presence or absence in the blood of certain protein materials called antigens A and B. These antigens are associated with the red corpuscles. We might picture a red corpuscle as a cell with these antigens attached to the outside surface. The four possible combinations result in the four blood groups illustrated in Fig. 52.

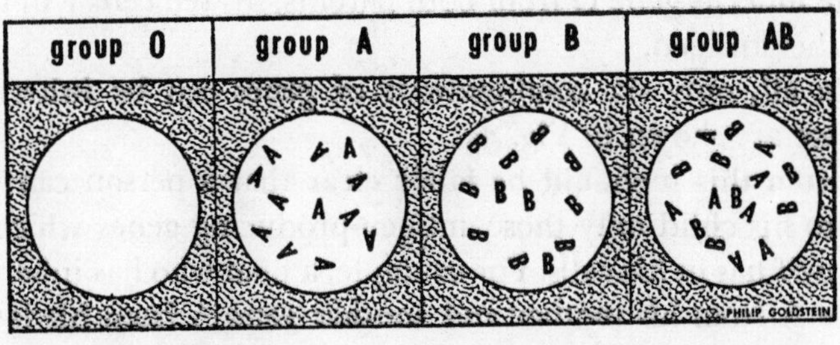

FIG. 52. Corpuscles of the four blood groups in the *ABO* series.

In a person who belongs to blood group A, the red corpuscles are "sprinkled" with the A antigen. The red corpuscles of a person in blood group B are "peppered" with the B antigen. People whose corpuscles bear both A and B antigens belong to blood group AB. Lastly, there are people whose corpuscles seem to be free of both antigens, and, since they have neither one, they are said to belong to blood group O.

As a result of Landsteiner's discovery, transfusions became relatively safe. A simple test now determines *in advance* whether or not the blood cells of a given donor will clump in the blood of the person awaiting transfusion. Only donors whose cells do not clump are permitted to donate blood. This testing procedure is known as blood typing, and usually, if the correct blood type is used, the transfusion can be carried out safely.

By 1925 it had been demonstrated clearly that the blood group to which a person belonged was an inherited trait. To be more exact, a person's blood group is determined, at the time of fertilization, by one gene from the sperm and one gene from the egg. If he inherits gene A from either parent, his red blood cells will have antigen A; if he inherits gene B from either parent, his red corpuscles will have antigen B; if he inherits gene A from one parent and gene B from the other parent, his red corpuscles will have both antigens; but if he inherits gene O from both parents, his red cells will have neither antigen.

In all, only six combinations of genes can be inherited. These are shown in Fig. 53.

From this it should be fairly clear that a person can pass on to his child only those antigen-producing genes which he himself has in his cells. For example, a man who has inherited gene A from his mother and another gene A from his father can pass only gene A on to his children. He cannot pass on gene O or gene B because he doesn't have any. And by the same token, *gene A must be present in every one of his chil-*

Parent No. 1 gives gene	*Parent No. 2 gives gene*	*Genes present in the fertilized egg*	*Antigens present in the red corpuscles*	*Blood Type*
O	O	OO	none	O
A	O	AO	A only	A
A	A	AA		
B	O	BO	B only	B
B	B	BB		
A	B	AB	A and B	AB

FIG. 53. The six combinations of genes which can be inherited in the *ABO* series.

dren. Fig. 54 shows which genes can and which genes cannot be passed on by persons of each of the blood groups.

This is the basis for the use of blood tests to determine parentage. The principle is simple enough. If the genes for blood group are inherited from the parents, then they must be found in the parents. Thus, for example, if a child has

Persons of blood type	*Must have genes*	*They CAN pass on*	*They CANNOT pass on*
O	OO	only gene O	gene A or B
AB	AB	either gene A or gene B	gene O
Persons of blood type	*May have genes*	*They CAN pass on*	*They CANNOT pass on*
A	AA or AO	either gene A or gene O	gene B
B	BB or BO	either gene B or gene O	gene A

FIG. 54. Which genes can be passed on by persons of the four blood groups in the *ABO* series, and which genes cannot.

type-B blood, we can be pretty certain that one of his parents had type-B blood. Likewise, if a child has AB blood, we can be pretty sure that neither one of his parents had type-O blood. Or to put it another way, a man with O blood could not possibly father an AB child, because such a child must have received an A gene from one parent and a B gene from

Parent No. 1 belongs to blood group	*Parent No. 2 belongs to blood group*	*Reproduction by these parents*	
		Can produce children of these blood types	But cannot produce children of these blood types
O	O	O	A, B, AB
AB	AB	A, B, AB	O
A	A	A, O	B, AB
B	B	B, O	A, AB
A	O	A, O	B, AB
B	O	B, O	A, AB
AB	O	A, B	AB, O
AB	A	A, B, AB	O
AB	B	A, B, AB	O
A	B	A, B, AB, O	.....

FIG. 55. A given pair of parents can give rise to children of certain blood groups, but not of other groups.

the other, whereas a man with type-O blood can pass on only O genes. Fig. 55 summarizes the kinds of children which can and cannot be produced by various pairs of parents.

However a blood test of this type cannot prove that a given person *IS* the parent of a certain child. It can only say that he *COULD BE* the parent, or that he *COULD NOT BE* the parent. Consider the following situation and you will see why.

Suppose a mother of type-O blood has a child of type A. This means that the father contributed the A gene, since the mother has no A genes to pass along. So we know that the father carried A genes, which means that he was either blood group A or blood group AB. But when you consider the fact that there are about thirty-five or forty million men in the United States who carry A genes, you can realize immediately that the blood test cannot possibly prove which one of them is actually the father. Theoretically, any one of the millions of men with A or AB blood could have been the father, and no blood test yet devised can pinpoint *THE* one.

However, in the same case we can rule out as father any person of blood group B or O, since people of these blood groups do not carry A genes. To summarize, the blood test cannot say definitely who the father is, but it can rule out certain individuals by indicating that they could not possibly have passed on the genes found in the child.

Now, let us examine an actual case, and see how blood tests are used in practice. Here is how a hospital mix-up was resolved. On a certain day in a given hospital, two babies were born, both male. Shortly after the women returned home with their infants, Mrs. R discovered a tag marked K on her baby. The other mother, Mrs. K, insisted that she had her own baby and refused to exchange. The matter came to court where all concerned agreed to submit to blood tests, with the following results:

Mr. R—group O	Mr. K—group O
Mrs. R—group O	Mrs. K—group AB
Baby No. 1—group A	Baby No. 2—group O

Fig. 54 shows that a person belonging to blood group O can pass on to his children only O genes. Therefore, any child produced by Mr. and Mrs. R must belong to group O. This couple could not possibly produce a group-A baby.

On the other hand, though Mr. K gave each of his children an O gene, Mrs. K passed along either an A gene or a B gene. Therefore children of this couple must have the genetic make-up AO or BO and must belong to blood group A or B. A child of blood group O could not be born to this couple.

Since the R's could not have an A baby but could have an O baby, and since the K's could not have an O baby but could have an A baby, it was quite obvious that the babies were mixed up, and the court ordered an exchange.

Of course, not every case works out quite that nicely. Very often no definite decision can be reached. For example, in a case of bigamy, where a woman lived with two separate husbands, it became legally important to determine which man was actually the father of a child which had just been born. Blood tests gave these results:

Child—group A	Husband No. 1—group A
Mother—group B	Husband No. 2—group A

In this case the mother must be of genetic make up BO, and she contributed an O gene to the child. Therefore, the A gene in the child came from the father. Since both men belong to group A, either one of them could have contributed an A gene. Obviously, the ABO blood group test can never determine which man was the true father.

Fortunately, there are other series of genes for blood groups, so another test was tried, this time for the MN series. The results were as follows:

Child—group M	Husband No. 1—group M
Mother—group MN	Husband No. 2—group N

We know that a person of group M has received a gene for M from each parent. This immediately rules out husband

No. 2 as the father, since he belongs to group N and can only pass along to his children the gene for N antigen. But the child has no N genes. On the other hand, husband No. 1 can pass on gene M, and therefore he could be the father. The court used these blood tests as a guide in rendering a decision.

Thus, when the ABO series of genes fails to give information about possible parentage, the scientist can fall back on the MN series. And if this too fails, he can try the P series, or the Rh series. As additional blood factors are discovered, identification may become more positive. In fact, some believe that the blood factors of each individual person vary to the point that at some future time, when we have learned more about blood typing, we will be able to use it for identification purposes as we now use fingerprinting.

According to Dr. Alexander Weiner, who knows as much about blood typing as any living person, no two individuals (except identical twins) can have the same combination of blood groups. He pictures a red corpuscle to be studded all over its surface with many different antigens, as in Fig. 56.

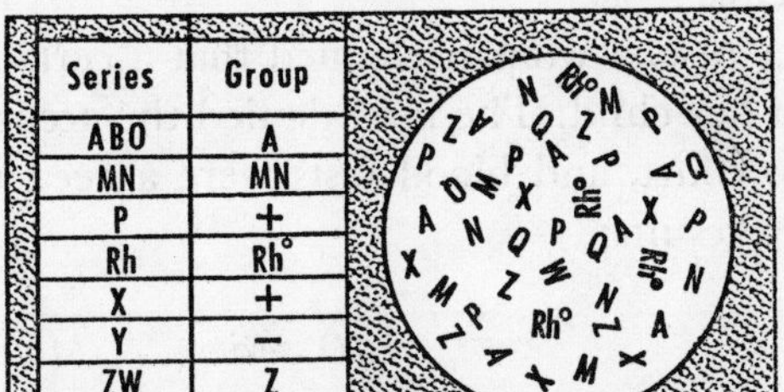

Fig. 56. An imaginary case—a red corpuscle of a person belonging to blood group *A, MN, P+, Rh°, X+, Y−, Z, Q. (Note—The X, Y, ZW, and QTL series are imaginary series, invented to illustrate the point that many different antigens stud the surface of a red corpuscle. Because of the great number of different combinations possible, we may someday be able to use blood grouping for positive identification of individuals as we use fingerprinting today.)*

At least twenty different factors are actually known at the present time so that over three million different combinations of blood groups are possible. As new blood factors are discovered, the number of combinations will increase tremendously.

Before closing this section, it will be of interest to analyze one or two more cases. Dr. Weiner describes one case in which the father of a pair of twin boys, aged five, *who did not resemble each other,* took his boys to a party. Here he noticed another boy of five who looked exactly like one of the twins. The father made inquiries, and it turned out that this boy had been born in the same hospital and on the same day as the twins. This raised the question of possible interchange, and blood tests were performed for the ABO series, the MN series, and the Rh series. It turned out that the dissimilar "twin" could not possibly be the son of the mother of the twins, while the other boy had exactly the same blood groups as the twin he resembled. So, after five years, an error was corrected, and the respective boys were restored to their rightful parents.

In another case, a woman claimed that a certain man was the father of her child. The man denied the accusation. The case came to court, and blood tests were agreed upon, with the following results:

	ABO series	*MN series*
The man	O	MN
The woman	A	N
The child	B	N

Based on the ABO series, one can readily see that neither the woman nor this man could have provided the B gene which the child carried, therefore another man, with B genes, must have been the father. However, a clever lawyer confused the jury by showing that the two tests "disagreed." He pointed out to the jury that, based on the ABO series, the man could not be the father, but, based on the MN series, the man could be the father. The two tests disagreed, he concluded, and therefore they could not possibly be reliable. He urged the jury to disregard the blood tests.

Evidently his reasoning was convincing, *even though it was wrong,* for the jury brought in a verdict in favor of the woman, to the effect that the man was the father of the child. Here is what seems to be a grave miscarriage of justice, brought about by the fact that the members of the jury did not understand the principles of heredity upon which blood tests are based.

CHAPTER 29

DEFECTIVE METABOLISM IN MAN

LIKE ALL OTHER living things, man is an intricate factory capable of carrying on the thousands of delicate chemical processes which produce the substances required by his cells. Control of these processes rests with the genes, which apparently synthesize the enzymes necessary to carry on the work.

Let us consider a simple, imaginary case. A person eats food which contains chemical substances called nutrients. One of these nutrients (we will call it substance A) is promptly converted into a new chemical (substance X) which nerve cells need in order to carry on their function. The ability to make this conversion is inherited, and it seems that the gene involved produces an enzyme (call it enzyme No. 1) which changes A into X. This is shown diagrammatically in Fig. 57.

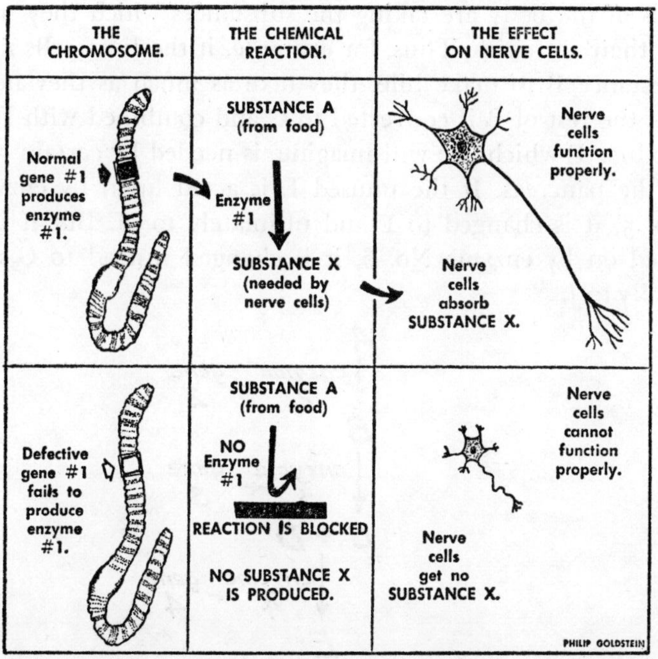

FIG. 57. An imaginary case—a defective gene may interfere with the proper functioning of some part of the body.

Now imagine that something goes wrong with gene No. 1, and it loses its power to produce enzyme No. 1. The chemical change from A to X can no longer take place because the enzyme is lacking, and since the nerve cells are not receiving X, they cannot function properly even though plenty of A is included in the diet. This is also diagramed in Fig. 57.

Naturally, things aren't quite that simple in the human body (or in the body of any living thing for that matter). Actually the chemical reactions run in complicated chains, as in Fig. 58. While these reactions are going on, the various cells of the body are taking the substances which they need for their activities. Thus, for example, if the liver cells need substance B to make bile, they take as much as they need. But the rest of B is converted to C and combined with D to produce E which, we will imagine, is needed by certain cells of the pancreas. If the unused E is acted upon by enzyme No. 5, it is changed to F and ultimately to H. But if it is acted on by enzyme No. 6, it is changed instead to G and finally to J.

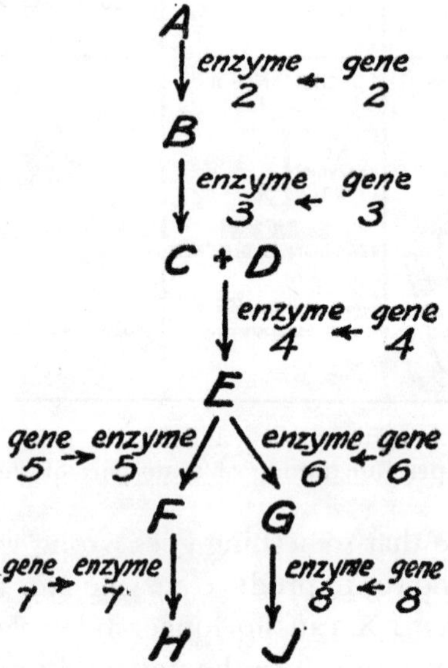

FIG. 58. An imaginary case—a complicated chain of genes.

Of course it is possible for any one of these genes to become defective as the result of a mutation. A change in gene No. 8 would not affect the entire process. Everything would be normal, except that substance G would not be converted to J. But suppose gene No. 2 breaks down. Now the whole process is blocked, and serious consequences might follow, since the defect is widespread in the body.

If this imaginary situation is understood, we are now ready to consider an actual case in the chemistry of the human body. Do not be alarmed by the chemical names of the substances involved, since the names themselves are unimportant. Consider them to be substances A, B, C, etc.

PHENYLALANINE is the name of one of the amino acids which man obtains from the proteins he eats. It is the jumping-off-place for a chain of chemical reactions essential in the metabolism of man. Connected with this chain of chemical reactions are the four human hereditary defects, listed below.

ALKAPTONURIA—in which the urine blackens on exposure to air

TYROSINOSIS—the condition where the body produces excess TYROSINE

PHENYLKETONURIA—a hereditary form of idiocy

ALBINISM—the condition where pigment is absent from skin and hair

At first glance, these four conditions seem to be quite independent of one another. They appear to have nothing in common except that they are all inherited as simple Mendelian recessives. Yet careful study of the chemistry of the body shows that each of the four conditions results from a block *at a different point in the same chain of chemical reactions,* beginning with phenylalanine. In the normal body metabolism, phenylalanine is the starting point for two inter-connected chains of reactions. Several of the steps in these two chains of reactions are known to be under the control of specific genes. The normal metabolism of phenylalanine is outlined in Fig. 59, with the genes indicated at the points at which they act.

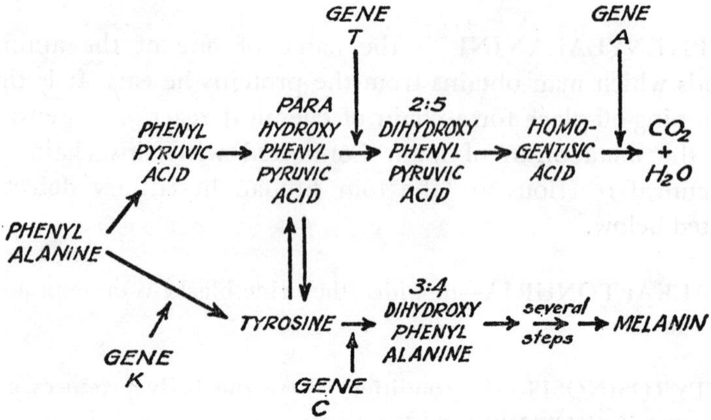

FIG. 59. An actual case—normal phenylalanine metabolism in man.

In the first chain of reactions, phenylalanine is gradually oxidized, in a series of steps, to carbon dioxide and water (see Fig. 60, Col. 1). Notice, that the change from parahydroxyphenylpyruvic acid to the next substance is controlled by gene T. Therefore, persons of genotype TT or Tt can carry out this reaction, since the dominant gene is present, and the necessary enzyme is produced. But a person who inherits only the recessive gene from each parent (tt) does not have the enzyme, and therefore the chemical change does not occur. As the parahydroxyphenylpyruvic acid piles up, it is converted into tyrosine, which is excreted in the urine. The person is afflicted with the condition known as tyrosinosis (see Fig. 60, Col. 3). Fortunately, there are no noticeable ill effects associated with this condition.

	Column 1	*Column 2*	*Column 3*
Chain of chemical reactions	Phenylalanine ↓ Phenyl Pyruvic Acid ↓ Para Hydroxy Phenyl Pyruvic Acid Gene T → ↓ 2:5 Hydroxy Phenyl Pyruvic Acid ↓ Homogentisic Acid Gene A → ↓ $CO_2 + H_2O$	Phenylalanine ↓ Phenyl Pyruvic Acid ↓ Para Hydroxy Phenyl Pyruvic Acid Gene T → ↓ 2:5 Hydroxy Phenyl Pyruvic Acid ↓ Homogentisic Acid gene a ↓ Out with urine No Enzyme reaction is blocked	Phenylalanine ↓ Phenyl Pyruvic Acid ↓ Para Hydroxy Phenyl Pyruvic Acid gene t → Tyrosine No Enzyme reaction is blocked ↓ Out with urine
Genes	Genes T and A both present in dominant form.	Gene T present in dominant form. Gene a present in recessive form only.	Gene t present in recessive form only.
Metabolism	Normal	Homogentisic acid excreted in urine.	Tyrosine excreted in urine.
Result	Normal	Alkaptonuria—Urine blackens on exposure to air.	Tyrosinosis.

FIG. 60. An actual case—how different faults in metabolism result from blocks at different points in the same chain of chemical reactions.

Similarly, the reaction can be blocked at the point where gene A normally acts. Persons of genotype aa cannot oxidize homogentisic acid, the last step in the chain. Such persons suffer from alkaptonuria, since the homogentisic acid (alkapton), which must be excreted in the urine, blackens upon exposure to air (see Fig. 60, Column 2). Again, no pathological changes are associated with this condition, except that in later life pigment may be deposited in the cartilage tissue of the body.

While the conditions described so far are interesting, they hardly interfere with the life or health of the sufferer. Blackening of urine may be unusual and inconvenient, but it does not shorten life or cause malfunction of body organs. However, suppose another "inborn error" of metabolism causes severe mental retardation, or some other condition that interferes with adjustment to the environment. What serious consequences could result!

Reference to Fig. 59 will show that the pigment *melanin* is produced from the amino acid *phenylalanine* by a series of chemical changes. Gene K controls one small step in this series, and Gene C controls another. This part of the total reaction is summarized in column 1 of Fig. 61.

If Gene C is present only in the inactive recessive form, the reaction will be blocked, producing a serious defect called *albinism*. Persons of genotype *cc* lack the enzyme necessary for carrying on the step shown in column 2 of Fig. 61, and as a result they cannot produce pigment in the skin, hair, or eyes. Such persons are known as albinos. They have very white skins; fine, silky, whitish hair; and whitish eyebrows and eyelashes. There is no color in the iris of the eye, and the eyes may appear pink or reddish. Naturally, albinos are at a decided disadvantage in life. They cannot tolerate the sun, and their eyes are usually weak and defective.

	Column 1	*Column 2*	*Column 3*
Chain of chemical reactions	Phenylalanine Gene K→ ↓ Tyrosine Gene C→ ↓ "Dopa" ↓ Several steps ↓ Melanin	Phenylalanine Gene K→ ↓ Tyrosine gene c No Enzyme Reaction is blocked	Phenylalanine gene k Phenyl Pyruvic Acid No Enzyme Reaction is blocked
Genes	Genes K and C both present in dominant form.	Gene K present in dominant form. Gene C present in recessive form only.	Gene K present in recessive form only.
Metabolism	Normal	Pigment melanin not produced.	Phenylalanine not changed to tyrosine, but to phenyl pyruvic acid which is excreted in the urine.
Results	Normal	Albinism—Absence of pigment in skin, hair, eyes, etc.	Phenylketonuria—A hereditary form of idiocy.

FIG. 61. An actual case—how different defects result when a chain of chemical reactions is blocked by different defective genes.

Albinism is found not only in man, but in many other animal species. The melanin-producing reaction is widespread through the animal kingdom, and hand-in-hand with the normal reaction, goes the defect caused by the inactive form of the gene. There are albino rabbits, albino mice, albino giraffes, albino deer, albino hippopotami, albino elephants, albino skunks, etc. It is quite likely that under natural conditions albinos of any species have trouble competing with their normally pigmented cousins.

Defects in Gene K lead to consequences even more serious. This gene produces an enzyme which controls the change of phenylalanine into tyrosine. Individuals of genotype *kk* cannot carry out this change. As a result, phenylalanine piles up in the body, and much of this is converted to phenylpyruvic acid which is excreted in the urine. But this is only a minor thing. Unfortunately, another result of this condition is a severe mental defect in which part of the cerebrum often shows signs of degeneration. The afflicted child may be handsome and well built, but his I.Q. may be as low as 25 or 30. Here then, is a form of hereditary idiocy which seems to result from faulty metabolism traced to a defective gene (see Fig. 61, Col. 3).

This "inborn error" was first identified when a mother brought two mentally retarded children to Dr. Asbjorn Folling, the Norwegian physician-chemist, with the complaint that her children had a peculiar odor. Investigation led Dr. Folling to conclude that the odor resulted from the presence of an unusual substance (phenylpyruvic acid or PPA) in the excretions of these children. When he found PPA in the urine of eight additional patients in local institutions for the mentally retarded, Dr. Folling began to suspect some relationship between PPA and retardation. He published his results in 1934, and soon thereafter his findings were confirmed by doctors all over the world. It was found that about 1% of all mental defectives suffered this condition, and the name *phenylketonuria* was applied to the type of mental retardation correlated with phenylpyruvic acid in the urine.

The problem of phenylketonuria has received much attention since that time. A simple test was devised to detect the presence of PPA in the urine of infants. The reagent is 10% ferric chloride added to fresh urine. If there is a gradual change in color to dark green followed by gradual fading, it denotes the presence of PPA in the urine. The test is so simple that it can be carried out by the mother in the form of a "diaper test". But testing is only half the battle. More important is the question, "What can be done if the test is found to be positive? Can the child be saved from idiocy?"

The only treatment so far devised is a dietary one. Restrict the phenylalanine in the diet. Eliminate all foods containing this amino acid on the theory that with less of this amino acid, the body will produce less PPA, and brain damage will be kept to a minimum. However, medical authorities point out that this treatment must be begun early, because the greatest damage to the brain occurs in the first few months of the baby's life. Therefore early testing is imperative. They recommend that every infant be tested within three weeks after birth, and then periodically for two or three years.

Apparently phenylketonuria is a deficiency disease caused by *the lack of something in the body*. Because the body fails to get the missing ingredient, some physiological process cannot occur and trouble results. Phenylketonuria resembles other deficiency diseases in this respect. Thus, the lack of insulin causes diabetes; the lack of ascorbic acid causes scurvy; the lack of the enzyme produced by gene K (Fig. 61, column 3) causes phenylketonuria. But we have learned to control diabetes and scurvy by providing the missing ingredient. Perhaps some future biochemical research will provide the enzyme lacking in phenylketonuria, and thus provide the means for overcoming another "inborn error of metabolism" resulting from the inheritance of a defective gene.

CHAPTER 30

THE INHERITANCE OF DISEASE

ARE DISEASES inherited? This question has been asked countless times, but, there is no single answer. One might say, "Yes," or "No," or "Maybe" and still be perfectly correct. It all depends on how you define the word disease, and what disease you are talking about.

Many diseases are caused by germs, or microbes. And a microbe is a complete living organism in its own right. It may be small, even invisible to the naked eye, but it is a complete living thing, with genes of its own and a heredity all its own. Something in its heredity gives the microbe the power to injure human tissues, or animal tissues, or plant tissues, as the case may be, and we then say that this particular microbe has caused a certain disease.

Can such a disease be inherited? It seems obvious that it cannot. We know that everything we inherit from our parents is contained in the genes of two cells—a sperm from the father and an egg from the mother. Can a microbe be carried in a human gene? When we know that a microbe is a complete organism with genes of its own, we realize that it must be considerably larger than any single gene, and therefore it could not possibly be carried in a human gene. So it should be fairly clear why diseases which are caused by microbes cannot be inherited.

But you may say, "I know of a case where a newborn baby had tuberculosis, or syphilis. How could a newborn child have the disease if it is not inherited?" The answer still remains that germ-caused diseases cannot be inherited, *but they can be caught very early in life.* The newborn child, in close contact with a tubercular mother, is constantly exposed to the germ, and, unless proper precautions are taken, the bacterium which causes the disease soon infects the child. In the case of syphilis the infection may take hold even earlier—even before the child is born. It is known that the spirochete which causes syphilis can penetrate through the placenta while the child is still developing within the body of the mother. In this case the child may be born with congenital syphilis. Yet the disease was not inherited. The egg and sperm which joined in fertilization were free of the disease, and proper medical attention could have prevented the infection from reaching the child.

Let us now consider another kind of disease, that which is caused by *the lack of something in the body* rather than by the entrance of a foreign organism. These are called deficiency diseases, and they are caused by what the body does *NOT* get. Examples of such conditions are the kind of anemia caused by the lack of iron in the diet; rickets, which is caused by the lack of vitamin D, or calcium, or phosphorus; goiter, which results from lack of iodine; and any one of a host of other similar diseases.

It seems clear that deficiency diseases are not inherited. You don't inherit scurvy, or pellagra, or iron-deficiency anemia. But in a sense, heredity is involved. Your genes determine that you *can* develop scurvy if your diet does not provide ascorbic acid (vitamin C). To understand what this means, compare the effect of depriving two different kinds of animals of vitamin C. When rats are fed a diet deficient in vitamin C, they remain perfectly healthy, but if we treat guinea pigs in the same way, they soon become very ill with scurvy. This difference in reaction stems from a difference in genes. The genes of the rat endow the animal with the power to *manufacture* vitamin C from other things in its food. Thus, whether the diet of the rat contains vitamin C or not makes little difference to the well-being of the animal. The cells can make their own. But the guinea pig is not so fortunate. It lacks the hereditary power to make vitamin C, and therefore it is dependent on its diet to provide the needed chemical. If the diet fails to provide it, the cells do not function properly, and this shows itself as a deficiency disease.

In this sense, man is like the guinea pig. He, too, lacks the hereditary power to manufacture ascorbic acid, so he must get it from the food he eats or suffer the ravages of scurvy. The same is true of many other deficiency diseases. Where the genes give the power to manufacture a chemical needed in the life of the cells, no deficiency disease can develop. But where the genes do not provide the power to make an essential chemical, a deficiency disease will develop *IF the environment does not provide the essential substance.*

We are now ready to consider a third type of disease, in which a part of the body breaks down or ceases to function. For example, when certain cells in the pancreas stop producing insulin, the body develops diabetes. Or if the thyroid gland fails to produce thyroxin, cretinism or myxedema may develop. Here again, the disease itself is not inherited, but there may or may not be a hereditary basis for the physical condition which underlies the disease.

Consider these three possibilities for the failure of the pancreas to secrete sufficient insulin:

1. The genes which control the development of these cells were defective. Since the cells did not develop properly they were unable to produce the hormone. This would illustrate a truly inherited condition.

2. The genes were perfectly all right, but during the development of the embryo, some environmental factor prevented these cells from developing. Again the hormone was

not produced, but this would illustrate an environmental condition rather than a hereditary condition.

3. The genes were perfectly all right, and development proceeded properly. But later in life some other factor (a tumor, perhaps) destroyed the cells, and again it was impossible for the hormone to be produced. This, too, would illustrate an environmental factor rather than a hereditary one.

In all three of these cases, the person would suffer from diabetes. Outwardly the symptoms would be exactly the same, yet the basic cause underlying the disease is different in each case. It could be a hereditary cause, or it could be an environmental cause, or it could be a little of each. Not even a doctor could tell. Still, it should be pointed out that there is some statistical evidence that conditions such as diabetes do run in families, and they are therefore said to be "familial." Whether or not this proves that they are hereditary is still undecided.

On the other hand, there are several ailments which are known to be gene-controlled. The classic case is hemophilia, a disease in which the blood does not clot. It has been established pretty definitely that hemophilia is caused by a defective gene carried by the sex chromosome, and the mode of its inheritance has been described on pp. 93-94.

Several other blood defects are known to be hereditary. In sickle-cell anemia the hemoglobin in the red corpuscles is affected, and under certain conditions the cells change shape. It is caused by a defective autosomal gene, and if a person inherits two such genes, his red corpuscles change shape to a greater degree than do those of a person who inherited only one such gene. (See pages 130-134 for more information about sickle-cell anemia.)

In Cooley's anemia, the red corpuscles are defective in a different way. This disease, which is inherited as a simple recessive, is so serious that those afflicted usually die. Hemolytic icterus is a condition in which the spleen destroys the red corpuscles, causing anemia and jaundice. This condition is inherited as an autosomal dominant.

Huntington's chorea is a mental disorder which develops rather late in life. It seems to be carried as a simple dominant trait. By contrast, amaurotic family idiocy is a condition which kills in early infancy, and involves loss of muscular power, mental deterioration, blindness, and death. It is inherited as a Mendelian recessive. Blindness may also be caused by a condition known as retinitis pigmentosa, which is carried by a dominant gene. In this condition pigment is deposited along the blood vessels of the retina, first causing restricted vision and then blindness.

These are only a few of the human diseases which are known to be inherited. There are many others. As we make progress in our study of heredity, it is likely that we will find a hereditary basis for many other afflictions. And this leads to the question whether cancer, too, has a hereditary basis. This will be considered in the next section.

CHAPTER 31

IS CANCER HEREDITARY?

THE FACT that certain diseases have a basis in the genes leads to the possibility that cancer, too, may be inherited. But before we can discuss the genetics of cancer, we must understand what cancer is. A cancer is a growth. But it is a growth of cells of the body itself. These are not foreign invaders which get into the body and do damage but, rather, native cells of the body, gone berserk. The growth of cells in a normal body is controlled—presumably by the genes. But sometimes a group of cells gets out of hand and begins to grow without control. These cells increase rapidly, wildly; they destroy other cells; they infiltrate new parts of the body and create havoc in these new locations. Ultimately they cause death. This is the nature of cancer, and it can happen to any of the many cell types of which the body is made.

There are many questions to be answered here. What causes apparently normal cells suddenly to go wild? If the genes control normal growth and development, why is this control suddenly lost? Is this loss of control also gene-controlled? In other words, is cancer inherited? The answers to these questions are not clear, but we may get some insight into them if we consider a few animal experiments which have been performed.

Let us consider the experiments with fish carried out at the American Museum of Natural History by Dr. Myron Gordon. He found that spontaneous cancers could be induced in tropical fish by mating certain species of swordtails with platys. This is curious, because the cancer does not appear in either the swordtails or in the platys alone. It occurs only in the hybrids which result from the mating. This is interpreted to mean that the genes of these two species complement each other to produce the conditions necessary for cancer to grow.

Next, let us turn our attention to the pedigreed mice which are being bred at the Memorial Laboratory in Bar Harbor, Maine. These mice have been inbred for many generations, so that those of any given strain are genetically as similar as identical twins can be. Here, they have one strain which they call their DBA strain. Every mouse in this group is an even dilute brown in color, and in generation after generation 90 per cent of the females develop a certain kind of breast cancer. At the same time, mice of the AKR strain, which is white in color, are almost certain to die of leukemia, a type of cancer of the blood. But in another white strain, the ST strain, only 1 per cent will get leukemia, and only 3 per cent will get lung or breast cancer. Still another white strain (A) shows a high incidence of breast and lung cancers.

This seems to indicate a definite hereditary basis for cancer. If cancer strikes with regularity in certain strains of mice and not in other strains, isn't it likely that there is a genetic difference among these strains? Isn't it probable that cancer is hereditary? Before reaching any conclusion, however, consider another experiment.

The newborn young of a DBA-strain female mouse were taken away from their real mother and nursed by a foster mother of a low-cancer strain. It was pointed out above that in the DBA strain 90 per cent of the females spontaneously develop breast cancers, but the mice which were suckled by a foster mother were relatively free of cancer. How is this possible, if cancer is gene-controlled? The suckling mother in no way changed the genes, yet the children were free of cancer. It was concluded logically that there was something in the milk of the DBA mice which triggered the cancerous growth. To test this, a reverse experiment was tried, in which the newborn infants of a low-cancer mother were suckled by high-cancer females. If the genes alone were responsible, very few of these children should have developed cancer. But actually, a larger number showed cancers when they grew up. So the milk of the high-cancer strain must contain something which stimulates the development of cancer of the breast. The factor in the milk which had this effect behaves like a virus. But is it? We don't know. Furthermore, this factor exists only in mice. No such milk factor has been found in any other animal. However there is evidence to implicate viruses in other forms of cancer.

Ordinarily, viruses take over the genetic machinery of the cell, and "order" the cell to manufacture component parts of new viruses. Shortly, many new viruses are assembled in the cell, the cell bursts, and the new viruses are freed to infect other cells. In these cases the host cell is destroyed.

But the mumps virus causes enlargement of the parotid glands without destroying the cells. A virus is responsible for the common wart. Certain mid-western cottontail rabbits develop papillomas (warty growths) that can be transmitted from rabbit to rabbit by injecting a little papilloma virus.

While these papillomas themselves are not malignant, cancer often develops in animals injected with the papilloma virus. There is no proof that the virus actually causes the cancer. In fact, the virus can never be detected in the cancer itself. But this leads up to another fascinating form of virus behavior.

Certain viruses enter a cell and disappear so completely that their presence can no longer be detected. The virus genes have become part of the genetic material of the host cell, and they are transmitted from cell generation to cell generation like any other group of genes. Their presence may confer new traits on the cell. Thus, the diphtheria bacillus can produce its virulent toxin *only when infected with a certain hidden virus.*

The virus of *herpes simplex* may remain quiet for long periods without causing any noticeable effect. But if the delicate host-virus balance is upset by exposure to strong sunlight, by fever, by severe emotional upset, a rash of "fever blisters" suddenly advertises its presence.

Similarly, certain viruses that infect bacteria are known to disappear into the genetic material of their bacterial hosts. They are transmitted generation after generation like any other bacterial genes. Suddenly something sets them off—certain chemicals, X-rays, ultra-violet irradiation, . . . They take over the cell mechanisms, reproduce, and burst the cell. Notice—these are the very same agents that are ordinarily considered to be carcinogenic (cancer producing).

Recently a form of leukemia in mice was traced to a virus. This form of leukemia can be produced in a certain percentage of susceptible mice by exposing them to X-rays. Can this mean that the X-rays activated a latent virus which in turn caused leukemia?

Biologists are beginning to lean towards the view that viruses are involved in at least *some* forms of cancer. One straw in the wind is the award of a Nobel Prize in 1966 to Dr. Francis Peyton Rous of Rockefeller University *"for his discovery of tumor-inducing viruses."* The interesting thing about this 1966 award is that Dr. Rous made his discovery about 55 years earlier, in 1911.

In 1909 a farmer brought to Dr. Rous a Plymouth Rock hen with a tumor in its breast muscles. Dr. Rous found the tumor to be a *sarcoma;* a form of cancer. He was able to transplant it from one fowl to another for many generations.

In 1911, he ground up some of the tumor into a soup and filtered it carefully to remove all cells and bacteria. He injected the clear liquid into nine healthy, young Plymouth Rock chickens. *One of them developed the typical sarcoma.* The new growth could then be transplanted in the usual way. Later, Dr. Rous found that warming the clear soup made it much more effective. In one experiment ten birds were injected with the warmed liquid, and *all ten developed tumors.*

Evidently, in the case of the *Rous sarcoma* (as this tumor came to be called) there was some "mysterious substance" in the clear fluid that transmitted the cancer. The "mysterious substance" has since been identified as a virus. Unfortunately, the same techniques failed to produce evidence of viruses in any mammalian or human tumor, and many biologists moved away from the notion that viruses are related to cancer. However, the pendulum seems to be swinging back, witness the honor bestowed on Dr. Peyton Rous in 1966.

We are still far from the answer. But it is tempting to speculate that latent viruses *hidden in the genetic material* of the cells are launched into "wild" activity by exposure to carcinogenic agents. This brings us back to our original question. Suppose the virus *IS* part of the genetic material. Does this mean that cancer is inherited?

There is an even more intriguing question to be considered and weighed. On page 25 it was suggested that genes might be "turned on" or "turned off" by other genes. Is it possible that controlled growth results when certain genes are "turned off" at the right moment when further growth of particular cells is unnecessary? Can it be that uncontrolled growth of certain cells (cancer) results when the genetic mechanism slips up and "forgets to turn off" the genes that control the growth of those cells?

CHAPTER 32

PHYSIOLOGICAL GENETICS—HOW GENES AFFECT DEVELOPMENT

WE KNOW that every many-celled animal begins life as a single cell—the fertilized egg. We also know that no matter what kind of animal it is, its egg begins to develop by the process called CLEAVAGE (see Fig. 62). In this process the egg divides rapidly, so that the original one-celled egg soon becomes a ball of many small cells, all more or less similar in appearance. One side of this ball folds in, and now there are two layers of cells. A third layer soon develops between the original two, but up to this point there are no special organs or organ systems. The body of every animal consists of these three layers of undifferentiated cells, which are known as the primary germ layers.

Soon the process called differentiation begins. The cells start to take on special forms—nerve cells, skin cells, blood cells, muscle cells, etc. These special cells and tissues arrange themselves into complex organs—heart, lungs, bones, intestines, kidneys, as the case may be; and soon an animal develops, fully equipped with a set of organ systems to enable it to live in the environment to which it is suited. At this stage it is born or hatched, and, following a period of growth, it becomes an adult animal, ready to create the next generation.

Cleavage, differentiation, and growth are as characteristic of a starfish egg as they are of a frog egg, or a bird egg, or a

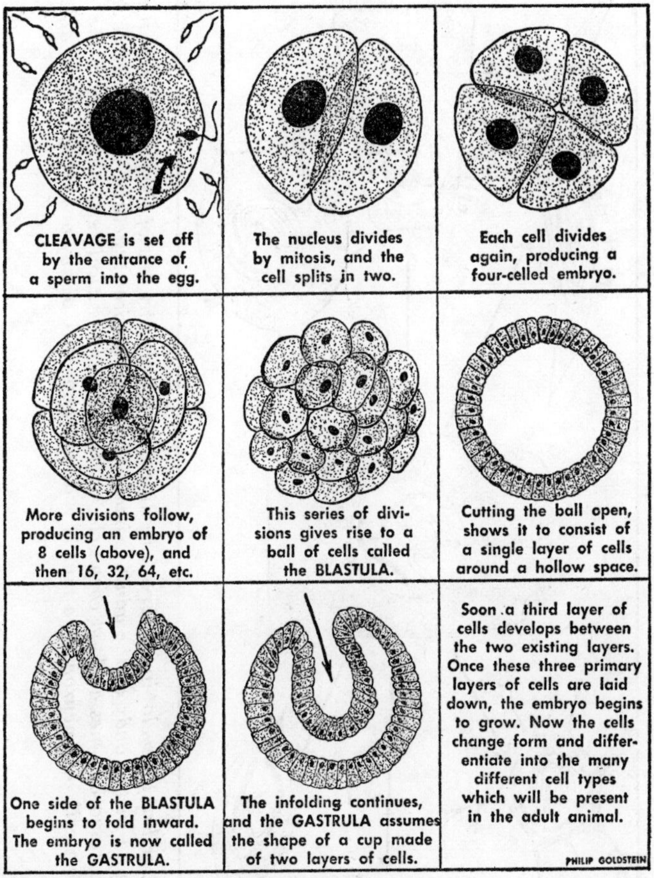

FIG. 62. Cleavage—the early development of a one-celled fertilized egg into a many-celled embryo.

human egg. Yet we may feel certain indeed that the starfish egg will infallibly become a starfish, and that the human egg will become a human baby in the normal course of events. Why can we feel so sure of this?

The answer lies with the genes. The pattern of development is an inherited pattern, controlled by gene action. The fact that development is fundamentally similar in all animals

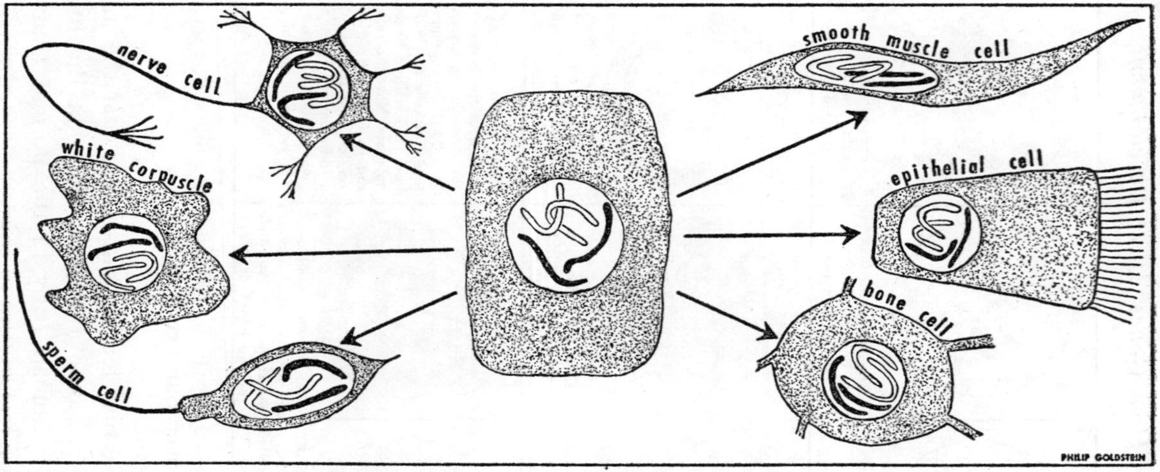

FIG. 63. A mystery. *A cell of a developing embryo can differentiate into any one of the many different types of cells which are found in the body of the adult animal. The physiological geneticist would like to know why one cell of the embryo becomes a blood corpuscle, while another identical cell, with exactly the same genes, becomes a sperm cell, or a neuron, or a muscle cell, or a bone cell, or any one of a hundred different kinds of cells.*

would seem to indicate that they all have certain genes in common, probably obtained from some ancient common ancestor from which they all stem.

Yes, the pattern of development of an animal is controlled by the genes. And here is a whole field for the geneticist to explore. How do genes control development? Why is it that similar cells with identical gene complements can differentiate into such varied types as nerve cells, blood cells, epithelial cells, bone cells, etc. (see Fig. 63)? At what point in development does a certain gene or group of genes take command of the changes which are to occur? These are among the questions which the physiological geneticist raises, but only a very tiny start has been made in the direction of finding some answers.

CHAPTER 33

CHAINS OF DEVELOPMENT—WITH GENES AS SPARK PLUGS

SPEAKING very generally, development appears to be like a series of chain reactions. One step sets off the next, which, in turn, sets off the next, etc. Genes are believed to control certain links in these chains of developmental processes. A gene might act as the spark plug which begins a chain of events. For example, a given gene may initiate the development of a particular group of cells. The cells, in turn, produce enzymes or hormones which regulate or alter the development of cells elsewhere in the body. Or, a defective gene may cause poor blood development. When circulation begins in the developing embryo, insufficient oxygen may reach the cells which develop into the backbone. These cells therefore divide less rapidly, and the result may be defective vertebrae.

In order to tie a particular gene to a particular phase of development, the geneticist relies largely on defects in development which crop up from time to time. When a defect is discovered, the animal is bred to normal animals in the regular Mendelian way to find out if the defect is hereditary and how it is inherited. Suppose a tailless mouse appears in normal stock, and the tailless condition is found to be inherited as a simple recessive. This means to the physiological geneticist that every mouse with gene T has a tail, whereas every mouse which has only the allelic gene t has no tail. He can assume from this that gene T must have something to do with the development of the tail. He might even be tempted to conclude that gene T controls the developmental process which leads to normal tail formation. However, just as there may be many reasons why a car refuses to start, there may also be many different reasons why the tail did not start to develop.

A. Mice without Tails

In an actual experiment with a certain strain of mice, it was found that the absence of a tail was inherited as a simple Mendelian recessive. But, peculiarly enough, in this particular strain of mice, lack of a tail was always accompanied by the absence of kidneys. Although such tailless mice lived and developed in the uterus of the mother, they died immediately after birth.

Careful study of the embryos of normal mice revealed that when the backbone was formed, kidney tubes developed alongside of the rear portion of the backbone. These kidney tubes soon developed into kidneys. On the other hand, in the tailless mice, formation of the rear part of the backbone was defective, and kidney tubes failed to develop. Therefore there were no kidneys in these mice.

It would seem that the "gene for tail" has something to do with the development of the rear portion of the backbone (including the tail). Once this begins, the rest of the chain of development follows. The presence of this portion of the backbone creates the conditions which stimulate the kidney tubes to form, and eventually kidneys develop. But in the tailless mice, the defective "gene for tail" could not play its part in the development of the backbone, and without the presence of the rear part of the spine, kidney tubes were not stimulated to develop. Result—no kidneys, and death.

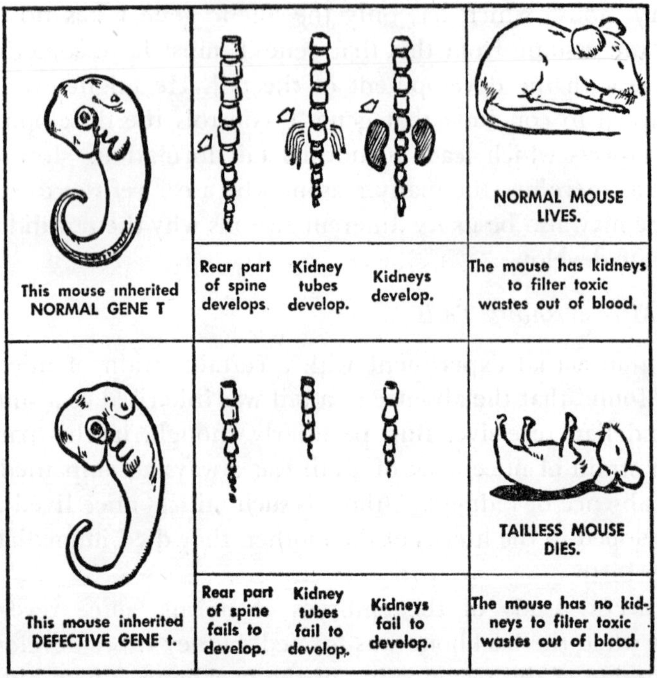

FIG. 64. Faulty development of the kidneys, and resulting death, is traced back to a defective gene—the gene for taillessness.

Here is a case where faulty development has been traced back to a single faulty gene. Fig. 64 illustrates this in diagrammatic form.

B. Killing Colors—Gray Lethal in Rats

Geneticists have discovered a number of cases where a certain color is lethal (see Sec. 13). But what relationship is there between color and ability to live? True, it is clear that albino corn cannot survive, because the lack of chlorophyll makes it unable to manufacture food. But why should a black or a hybrid yellow mouse survive, while a pure yellow mouse dies before it is born?

A partial answer has been worked out, in the case of rats, for the characteristic known as "gray lethal." This particular kind of gray is dominant over the wild type, but when two of the hybrids are crossed, the F_2 ratio is 2 gray to 1 wild instead of the expected 3:1 ratio. Evidently pure gray is lethal, but why is this so?

Examination of the embryos shows a multitude of defects, such as a narrow windpipe which would make breathing difficult, a deformed mouth which would make suckling impossible, defective teeth, etc. As a matter of fact, there is faulty development of cartilage throughout the body, and many structures which depend on cartilage are defective. So it is not the color which kills but, rather, the faulty way in which various parts of the body develop. The gene which is labeled "gray lethal" in some way interferes with the proper development of cartilage, and thus an embryo which cannot survive is produced. However, what the gene does to cause this faulty development is not yet known.

C. Short-legged Dwarfs—Creeper Fowl

There is a certain variety of chicken known as the "creeper" fowl. Chickens of this breed are not normally proportioned. They have heads and bodies of normal size, but their legs and wings are much shorter than they should be. Because of their very short legs, these birds seem to creep along the ground, and from this characteristic the name "creeper" fowl is derived.

Study of the creeper trait by the regular Mendelian method showed it to be inherited as a dominant trait, where pure dominant is lethal. In other words, a chicken inheriting two dominant genes for creeper is doomed to die before it even hatches from the egg, whereas a chicken receiving only one such gene hatches into a creeper. Normal birds have two genes for the recessive allele.

But why do creepers develop this way? What does the gene

do to make the creepers different from normal birds? To answer such questions, physiological geneticists began a careful study of the developing embryos of creepers and normal chickens. It was soon found that the pure creepers showed very early evidence of poor growth. Structures such as the head, the eyes, and the ears, which develop early in normal chicks, failed to do so. They remained underdeveloped and small, as though something were holding them back. And then, after only four days of incubation, these embryos died.

The hybrid creepers, however, did not show signs of faulty development quite so early. They seemed to start right, but just at the time when normal chicks begin to develop wings and legs, something went wrong. So, the growth of these parts was retarded, and the chicks hatched out with shortened wings and legs—creeper fowl.

This seems to indicate that the so-called "creeper" gene really retards the growth of whatever part of the body happens to be undergoing the greatest amount of growth at the moment when the gene acts. Thus, when the gene acted very early, it retarded growth of the head, eyes, ears, and other parts which are vital to life. But when it acted a little later, after these parts had already grown, it only affected the wings and leg bones which were just beginning to develop.

Perhaps someday our understanding of this will be carried a step further. Does the gene retard growth because in some way it cuts off circulation to certain parts, thus depriving them of the food and oxygen they need for growth? Or does the gene block some chemical reaction which is needed for growth? Or is there still another explanation?

CHAPTER 34

GENETICS OF MICROORGANISMS

IN RECENT YEARS genticists have turned their attention to microorganisms such as bacteria, yeasts, and molds. This may surprise you, for you may wonder whether simple organisms such as these inherit things. After all, some of them, such as bacteria, do not even seem to have a visible nucleus. How, then, can they have genes to pass on? What would geneticists want with such organisms?

But if you reflect for a moment, you will realize that they must have genes. When a scientist prepares a culture of diphtheria bacilli, he expects to find diphtheria bacilli growing in the test tube. And when he prepares a culture of wine yeasts, he expects to get wine yeasts. The very fact that we expect to find growing in the culture the same thing that we put into it shows that we believe the organisms have genes. Of course the new organisms are similar to the old because genes are transmitted during the process of reproduction.

Working with microorganisms has certain advantages. They are small, so they don't require much room. They reproduce rapidly, so the geneticist can study many generations. Many of them exist in the monoploid condition (with only one set of genes) so that any mutation which occurs is immediately visible, there being no second gene to mask the effect. They have comparatively simple needs and can be cultured easily. And of course, some of them play important parts in our lives.

Bacteria and viruses, for example, are known to cause many human diseases. Perhaps if we understood the heredity of these organisms we could better control the spread of the diseases. We know that certain diseases, which are with us all the time in comparatively mild form, will suddenly flare up and spread rapidly over the country, or even over the world. We then say we are having an epidemic. But why the sudden rapid spread? It has been suggested that this is tied up with a mutation in the disease organism. The mutation changed the organism in such a way that the resistance which man has built up over the years is no longer effective. Since there is little resistance to the new form, the disease spreads rapidly, and many people become sick.

Penicillium notatum is just an ordinary blue or green mold frequently found growing on fruits or cheese, yet it has a wonderful hereditary power. It has the inherited ability to produce the wonder drug, penicillin. Different strains of the mold vary in their ability to produce it. Some produce a great deal, some only a little. By tampering with the genes, scientists have been able to produce new strains of *Penicillium* which yield much more penicillin than did the original strains.

Some of the work being carried on with the mold *Neurospora* has already been described in chapter 27. The next section will tell a little about the genetics of yeasts, and will serve as a sample of the kind of work being done with microorganisms.

CHAPTER 35

BREAD AND BEER—THE HEREDITY OF YEASTS

WILD YEASTS have existed on earth since time immemorial. They can be found growing wherever sugar is available—in flowers, on fruits, on leaves. As they grow, yeasts carry on the process of fermentation, in which they break down the sugars for their own benefit. But in doing so, they produce alcohol and carbon dioxide. Long ago man learned to take advantage of this. Fermented fruit juices became alcoholic beverages such as beer and wine, while fermented doughs made fine breads and cakes. So, yeasts became the basis of two big industries, baking and brewing; and over the years special strains of yeasts were selected and cultivated because they were especially useful in these industries. In this way the standard strains of brewers' yeast and bakers' yeast were established.

Although yeasts have been cultivated for years, until recently little was known about their heredity. Actually, little could be learned about their heredity as long as it was believed that yeasts reproduce only asexually—by budding and spore formation. However, in 1918 two Czech scientists observed sexual reproduction in a common brewers' yeast, and in 1935 the life cycle of yeasts was worked out in detail by Winge in Denmark. He rediscovered, independently, the process of sexual reproduction in yeasts.

Winge showed that yeasts undergo a regular reproductive

cycle which includes both sexual and asexual reproduction. Yeast plants themselves can live in either the monoploid condition (with one set of genes) or in the diploid state (with two sets). In either form the yeast can bud normally and carry on fermentation. However, monoploid cells sometimes conjugate to form diploid cells, and diploid cells sometimes form spores. During spore formation, reduction division occurs, so that each spore is monoploid. The life cycle of a yeast is shown in Fig. 65.

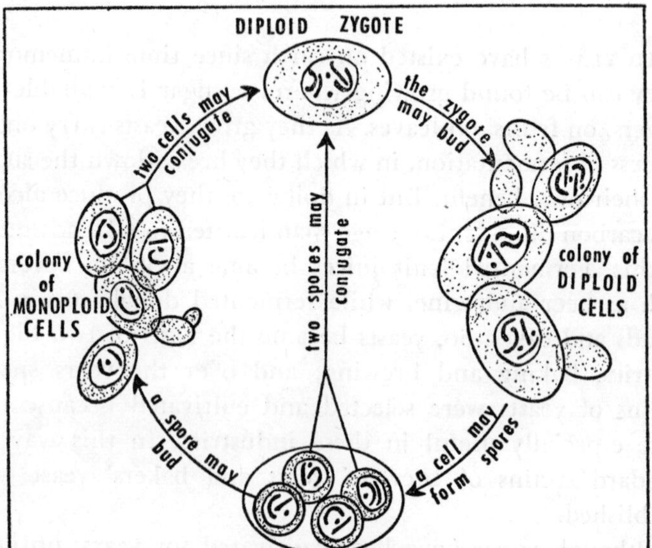

FIG. 65. The life cycle of a yeast.

Once it was established that yeasts conjugate, and once the techniques for bringing about conjugation were perfected, the heredity of yeast could be studied by the regular Mendelian method. For example, as explained on p. 47, there are some yeasts which can ferment a certain sugar, while other yeast varieties cannot. When a fermenting variety is crossed with a nonfermenter, the hybrid always has the ability to ferment the given sugar because it still has one gene for production of the necessary enzyme.

What happens when this hybrid reproduces? To find out, yeast geneticists allow the hybrid plants to form spores. Then they dissect the spore case, and isolate each of the four spores. Each spore is grown in a separate culture, and, as it germinates, fusion occurs, so that soon each spore has produced a colony of diploid cells. Half of the colonies are fermenters, and half are nonfermenters.

At first glance, this ratio seems wrong. We expect a 3:1 ratio when hybrids reproduce. However, careful study of Fig. 66 shows why, in yeasts, this ratio actually should be 1:1.

In making crosses such as the one described, a form of hybrid vigor is observed. The hybrids usually have greater fermenting power than either parent. This is easily explained. Suppose we have a solution containing equal amounts of two kinds of sugar—sucrose and melibiose. A variety of yeast which can ferment sucrose but not melibiose can act on only half the sugar in the solution. Likewise, a yeast which can ferment melibiose but not sucrose can act on half the sugar. But because the hybrid has inherited genes for both kinds of fermentation, it can act on all the sugar in the solution.

The ability to synthesize vitamins of the B complex is inherited in yeasts in the same way as is the ability to ferment sugars. In one experiment a strain of yeasts, which could make pyridoxine but not biotin or pantothenic acid, was crossed with another strain, which could make biotin and pantothenic acid but not pyridoxine. The hybrid was able to manufacture all three of these vitamins. Perhaps by crossing the proper strains, new hybrid yeasts will be developed with the power to synthesize all of the vitamins of the B complex.

Naturally such a hybrid can be expected to "breed true" only when it reproduces asexually—by budding. As soon as

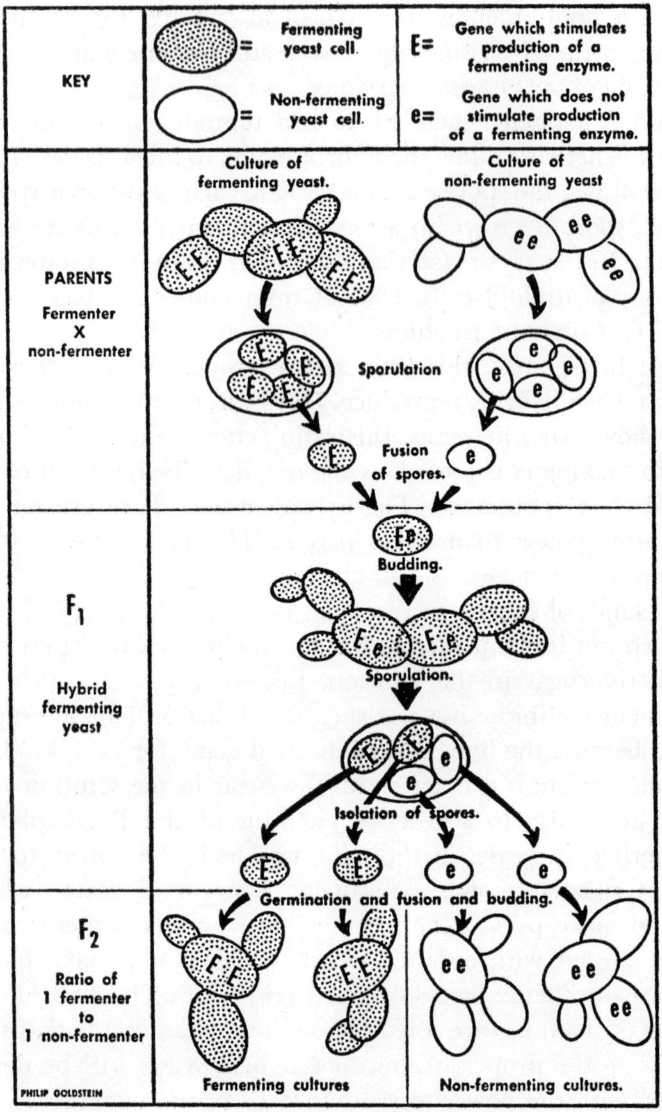

FIG. 66. How the ability to ferment a certain sugar is inherited in yeasts.

sporulation occurs, the dominant and recessive genes will segregate from each other, and only chance controls the recombination of genes.

Development of pure strains of yeasts is further complicated by the problem of mutations. Since the rate of reproduction in yeasts is so rapid, a single mutant cell may reproduce millions like itself in the space of a few hours, and it has frequently happened that an entire culture has gone bad because an undesirable mutant was able to grow faster than the desirable strain from which it originated.

Like all other organisms, yeasts are "blessed" with lethal genes. In one experiment only half of the spores in the spore case were able to germinate and develop into colonies. The others died because they inherited lethal genes which in some way interfered with the functioning of the cell.

CHAPTER 36

SEX AND HEREDITY IN BACTERIA

Escherichia coli (or *E. coli* for short) is a common bacterium of the human digestive system which has served as the subject of innumerable research projects. Like all bacteria, it was supposed to reproduce only by asexual means. That is why the scientific world was a bit startled in 1946 when Joshua Lederberg and Edward L. Tatum published a research paper entitled: *Gene Recombination in Escherichia coli.*

The words "gene recombination" implied that Mendel's method of genetic analysis had been used successfully with a bacterium. As explained in chapter 25, Mendel's method consists of crossing organisms having allelic forms of a trait, inbreeding the F_1 generation, and analyzing the gene recombinations in the F_2.

Obviously, this process cannot even begin unless there are opposite sexes to be mated. Therefore the Lederberg-Tatum report indicated that some form of sexual reproduction had occurred in *E. coli.* This represented a complete break through. It opened the door to a brand new field of genetic research—the field of *bacterial genetics.*

Actually, these experiments with *E. coli* were an outgrowth of the *Neurospora* work of Beadle and Tatum (Chap. 27). *Neurospora* has few nutritional requirements because it inherits the ability to synthesize nearly every vitamin and amino acid it needs. Beadle and Tatum, however, devised a way to reduce the synthetic powers of the mold by "knocking out" some of the genes on which these powers are based.

Bacteria are also great synthesizers. Like *Neurospora,* they too, inherit genes for the manufacture of nearly all the nutrients they need. Tatum decided it was time to find out whether the techniques used so successfully on *Neurospora* could be used to "knock out" some of the genes controlling synthesis in bacteria. The answer turned out to be a resounding "Yes!", and before long, he had produced several nutritionally deficient strains of *E. coli.*

In order to understand the classic experiment that grew out of this, let us consider the six nutrients listed in the table below which *E. coli* normally manufactures for itself. Presumably there is a gene underlying each synthesis.

GENE	CONTROLS THE SYNTHESIS OF	
B	The vitamin	BIOTIN
C	The amino acid	CYSTINE
L	The amino acid	LEUCINE
P	The amino acid	PHENYLALANINE
T	The vitamin	THIAMINE
TR	The amino acid	THREONINE

A normal strain of *E. coli* carries in its genetic make-up all six of these genes in active form (-B-C-L-P-T-TR-). Therefore it can synthesize all six of the listed nutrients. (Of course there are many other genes for the synthesis of other nutrients, but since we are interested only in these six at the moment, we will omit mention of the others.)

Now suppose that a wandering X-ray "knocks out" gene L. The genetic make-up of the mutant is -B-C-*l*-P-T-TR-. (The small *l* indicates the inactive or defective recessive form of the gene for synthesis of leucine.) This strain can still produce five of the nutrients under consideration. But with regard to leucine it is nutritionally deficient. It is said to be *leucine-less.* It cannot live unless the culture medium supplies it with leucine.

By exposure to radiation, followed by proper selection, other strains can be produced that are deficient in a variety of other syntheses. Some of them are deficient in one way only; others in two or three or more. Thus, a strain with genetic make-up -B-C-*l*-P-*t*-TR- is both *leucine-less* and *thiamine-less.* A strain of genotype -B-C-*l*-P-*t*-*tr*- is *leucine-less, thiamine-less* and *threonine-less.*

All this seems quite clear-cut. But mighty strange things happened when Lederberg and Tatum incubated some of these nutritionally deficient strains *together in the same culture tube.* For example, in one experiment, they began with two strains of *E. coli,* each deficient in three syntheses.

Strain Y-10—*thiamine-less, threonine-less, leucine-less.*
Strain Y-24—*phenylalanine-less, cystine-less, biotin-less.*

Each of these strains requires nutritional support. Strain Y-10 can grow only if supplied with thiamine, threonine and leucine. Strain Y-24 can exist only if supplied with phenylalanine, cystine, and biotin. Obviously, neither strain could possibly grow on a culture medium lacking in all six nutrients.

But wonder of wonders! When Y-10 and Y-24 were mixed and incubated together in the same culture tubes, colonies were found growing perfectly well in a medium which *did not supply a single one of the six crucial nutrients!* When these colonies were examined more closely, they were found to possess all six synthetic abilities: proving that their gene make-up was -B-C-L-P-T-TR-.

How was this possible? Not a single bacterium of this genotype had been put into the culture tube. The strains put in were

-B-C-l-P-t-tr- (Y-10)
and -b-c-L-p-T-TR- (Y-24)

The strain that came out was -B-C-L-P-T-TR-. Where did it come from?

There seemed to be only one logical answer. There must have been an exchange of genetic material; a recombination of some sort, in which one parent supplied three dominant genes (L, T, TR) and the other parent supplied the other three (B, C, P). Can you think of any other explanation?

Lederberg and Tatum were forced to the conclusion that they had demonstrated sexual reproduction in *E. coli.* This was the beginning of bacterial genetics, and the beginning of a brilliant career for Joshua Lederberg. He continued to make one discovery after another about bacterial heredity, and as he worked, a pattern began to emerge. The genes of *E. coli* were not scattered haphazardly about the cell in helter-skelter fashion. They were arranged in groups, comparable to the chromosomes of other organisms. Still, Lederberg could not say exactly where in a chromosome any particular gene was located, or even how many chromosomes there were in the cell.

However, help arrived from a different source. In 1952 Dr. William Hayes of London demonstrated that *E. coli* cells were of two types—male and female. Even though they looked quite similar on the outside, they behaved quite differently during the mating process. The experiment which led to this discovery is outlined in figure 67.

In this experiment, the male cells actually transferred some of their genetic material to the female cells. The female cells merely accepted the donated genes. This discovery opened the door for some unusual experiments performed by two French geneticists, Eli L. Wollman and François Jacob.

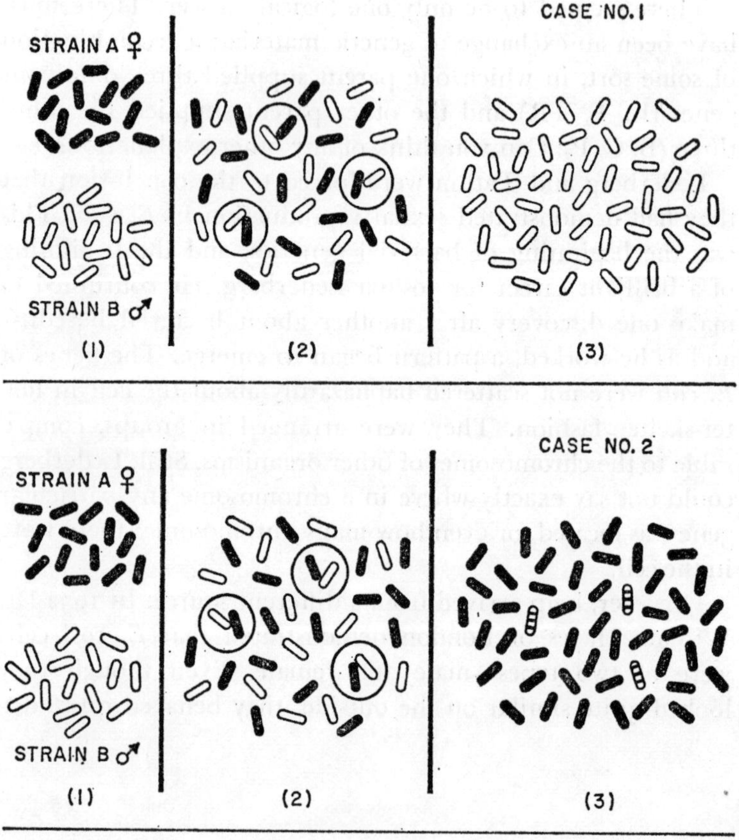

FIG. 67. The Hayes experiment with streptomycin-treated strains of bacteria.

Wollman and Jacob began a careful study of conjugation in *E. coli.* They discovered that the male and female cells remained joined for about half an hour, during which time the genes of the male cell slowly migrated into the female cell. (See Figure 68.) Following conjugation the cells separated and reproduced. *E. coli* cells with new combinations of traits (recombinants) appeared in the culture; combinations which did not exist in the original parental strains; recombinations like those seen in the earlier experiments of Lederberg and Tatum.

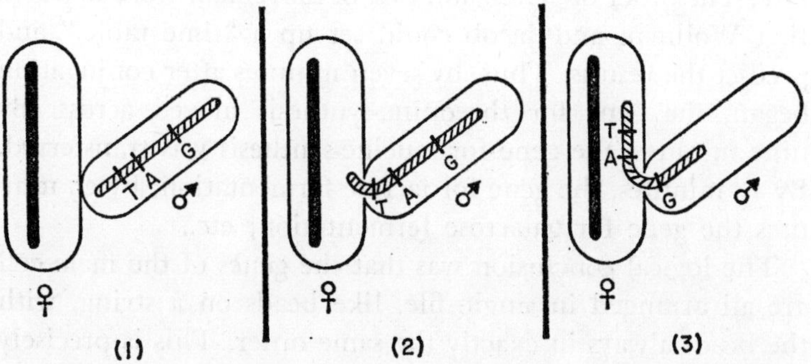

FIG. 68. During conjugation in *Escherichia coli* the chromosome of the male cell gradually migrates into the female cell.

But a thought struck the two experimenters. What would happen if the conjugating cells were broken apart suddenly, five minutes after the process began; or ten minutes; or fifteen . . . ? They devised an ingenious experiment to make the test. They incubated together some male and female cells. Then at a given moment, they removed a part of the incubating culture and whirled it furiously in an electric blender. As the conjugating cells were spun around, they broke apart at whatever stage of conjugation they had reached.

Suppose the whirling interrupted conjugation exactly five minutes after it began. *Only those genes that are transferred in the first five minutes of conjugation could possibly reach the female.* But if blending began after ten minutes of conjugation, a greater amount of genetic material could be transferred. Even more genes could move across if fifteen minutes elapsed; and so forth. Thus, by blending at different intervals, Wollman and Jacob could control the total transfer of genes.

They found that the genes always seemed to move from male to female *in exactly the same order, and at a very definite rate.* The order of genes, and rate of movement were so exact, that Wollman and Jacob could set up a "time table," and predict the results. Thus, by seven minutes after conjugation began, the gene for threonine-synthesis moved across. By nine minutes, the gene for leucine-synthesis was transferred. By 18 minutes, the gene for lactose fermentation; by 25 minutes the gene for galactose fermentation; etc..

The logical conclusion was that the genes of the male cell are all arranged in single file, like beads on a string, with the beads always in exactly the same order. This is precisely the way a chromosome is usually pictured. During conjugation, the male chromosome moves steadily into the female cell. Since the genes are always arranged in the same order, they must always enter the female in that sequence.

An analogy may be helpful. Think of an imaginary worm which has on its back a sequence of colored spots as brilliant as the colors of a rainbow; and in the same order—red, orange, yellow, green, blue, indigo, violet. We watch this imaginary worm eat his way into an imaginary apple. (And we compare it to the gradual entry of the male chromosome into the female *E. coli* cell.)

The worm eats at a steady pace. One minute after he began, his head is buried in the apple, but all the colored spots are still outside. At the end of three minutes, he is so far buried in the apple that the red spot no longer shows. After 5 minutes, the orange spot has also disappeared. And so on, until the violet spot disappears 25 minutes after he began work.

But if we so desire, we can interrupt the worm at any moment during his imaginary meal (just as Wollman and Jacob interrupted the bacterial conjugation). With a pair of sharp scissors we can snip off that part of the worm which is still outside. If this is done very early, only a few of the colored spots are left in the apple. The longer we delay the snipping, the more of the worm, and the more colored spots get into the apple.

Bacterial conjugation is quite similar. The longer the male and female cells are allowed to remain in contact, the more male genes can be transferred to the female. In nature, the conjugating cells can break apart at any time during conjugation. But even if conjugation lasts for a short time, and only a small part of the males genes entered the female, some sort of genetic recombination has occurred.

CHAPTER 37

THE GENETIC MATERIAL: PROTEIN OR NUCLEIC ACID?

ACCORDING TO the gene theory (Chap. 25) heredity is transmitted from generation to generation by a set of discrete units called genes. These genes are arranged in single file along the chromosomes, with each gene occupying a very specific spot (locus). Analysis of chromosomes (chap. 18) indicates that they are made of proteins and nucleic acids. This leads to the logical question, *"Is the genetic material in the protein part of the chromosome or in the nucleic acid part?"*

To function as genetic material, a substance must satisfy certain basic characteristics. Its molecules must possess unusual and remarkable properties, particularly those listed below.

1. VARIATION: genetic material must be capable of almost infinite variation in structure; variation great enough to account for all the different genes that exist in the plants and animals of the world.
2. REPLICATION: genetic material must be capable of reproducing itself with precision and exactness if identical genes are to be transmitted unchanged from one generation to the next.
3. MUTATION: genetic material must be capable of slight change from time to time to allow for the possibility of mutation.

4. SYNTHESIS: genetic material must be able to cause the synthesis of all the other kinds of molecules necessary to exercise control over the development and metabolism of the organism.

From the beginning, scientists assumed that genes must be protein materials. The exactness of a gene might be explained on the basis that the amino acid chain of a given protein is always specific and exact. The infinite number of different genes might be explained by the fact that the arrangement of amino acids in the protein chain can vary in an infinite number of ways. And a slight accidental change in amino acid arrangement might easily change the gene and so produce a mutation.

Yes, proteins could fill the bill very nicely. If only one could explain how a chain of amino acids can reproduce itself; or how a protein can cause the synthesis of other proteins in the body of a living thing.

The early geneticists were so certain of the protein nature of genes, that they practically ignored the nucleic acids. These acids were discovered shortly after Mendel's paper was written. In the years that followed, they were examined and analyzed many times. But their chemical nature did not seem to allow for the great variability required to account for so many different genes. Without great variability the nucleic acids did not seem to qualify as genetic material. At least that is what geneticists thought.

However, in recent years there has been a complete turn about. It is now universally agreed that the hereditary material resides in the nucleic acids rather than in the proteins. The next few chapters will present some of the evidence that brought about this radical change of view. Let us begin with the story of *"the transforming principle."*

CHAPTER 38

TRANSFORMATION: HEREDITY FROM THE DEAD

ONE OF THE commonest causes of pneumonia is the family of microbes called the pneumococci (*Diplococcus pneumoniae*). Because of their medical importance, these germs have been studied quite carefully, and a great deal of information has piled up about their characteristics.

Usually, the diplococcus appears as two spherical cells (cocci) within a surrounding capsule. The capsule contains certain polysaccharides (complex sugars) that separate the family into different types or strains. Even though these various types look very much alike on the outside, there are important differences. Each type of pneumococcus apparently produces a different kind of sugar in its capsule, and a slightly different form of pneumonia. This gave the doctors quite a headache because it was impossible to develop a single effective serum for all cases. A serum developed against Type II pneumococci (for example) did not protect against Type III.

Although in those days bacteria were considered to be too simple to have genes, the pneumococci seemed to maintain their traits very well. Type II germs grown in laboratory culture for long periods of time remained Type II, with all the traits of Type II. Type III remained Type III.

In laboratory cultures, disease-causing pneumococci usually formed glistening white colonies, with a glossy smooth surface. Occasionally something happened, and a "freak" colony appeared, having an uneven, rough surface. Pneumococci from such *rough* colonies did not seem to have the capsule that is so typical of their cousins from *smooth* colonies. But more surprising; and far more important—*the cells from a rough colony could no longer cause disease.* Evidently, the ability to cause disease is related to the capsule and its polysaccharides.

All this information can be summarized by comparing the characteristics of Pneumococcus Type III *Smooth* and Type II *Rough.* Type III (S) is a disease producing strain whose cells are coated by the typical capsule. It forms glistening, smooth colonies. Type II (R) is a non-virulent strain. Its cells lack the capsule, and it forms colonies with a rough, uneven surface. This was about the state of knowledge in 1928 when an English scientist named Frederick Griffith made a series of observations on mice. Some of his preliminary experiments are listed in the table below.

Experiment #	*Type of pneumococci injected into mice*	*Effect on the mice*	*Why this result was expected*
1	Living Type II (R)	HARMLESS! The mice remained healthy	R-type pneumococci do not cause disease.
2	Living Type III (S)	VIRULENT! The mice died of pneumonia	S-type pneumococci are virulent disease producers.
3	Heat-killed Type III (S)	HARMLESS! The mice remained healthy	Dead pneumococci do not cause disease.

So far these are run-of-the-mill experiments, with nothing to set them apart from thousands of similar experiments in which things turned out just about as was expected. But now comes the "crazy" part. For some unexplained reason Dr. Griffith decided to inject another group of mice *with two cultures simultaneously.* Into these mice went some *harmless living type II* (*R*) cells, plus some *harmless heat-killed type III* (*S*) cells (see #1 and #3 in table above).

Since both of these are perfectly harmless cultures, one would hardly expect the mixture to cause trouble. But much to Dr. Griffith's surprise, some of the mice began to die of pneumonia. And more startling still, although not a single living III (S) cell had been injected into the mice, *living III* (*S*) *cells were recovered from the blood of the sick mice.* Where could they have come from?

Contamination with living III (S) cells was quickly ruled out, leaving a startling idea. Did the dead III (S) cells suddenly come back to life? What an unlikely thought. This had never happened in any other experiment involving heat-killed III (S) cells. But was there another answer possible?

Perhaps the living II (R) cells were altered in some way by something from the bodies of their dead III (S) cousins. Did something from the dead III (S) cells enter the living II (R) cells and change their harmless nature? Were the harmless II (R) cells *transformed* into the virulent III (S) type?

This hypothesis sounded incredible. But soon it was confirmed many times in many laboratories. The harmless II (R) cells *COULD BE TRANSFORMED* into vicious killers merely by mixing them with the dead bodies of heat-killed III (S) cells. The transformation could be brought about in the bodies of living mice (as Griffith first showed) or even in test tubes in the laboratory.

But of course, nobody could pin down the something from the dead III (S) cells that entered the living II (R) cells to transform them. The name "*transforming principle*" was assigned to it, and it was left undefined. About 15 years later a group of American scientists headed by Dr. Oswald T. Avery went back to the problem of transformation. Three years of painstaking work enabled them to prove that "the transforming principle" was really a nucleic acid—*deoxyribonucleic acid (DNA)* to be exact.

Dr. Avery's group was very cautious in interpreting the results. But we can throw caution to the winds and look at it in this way:

Heating the III (S) culture killed the cells and caused them to release some of their DNA. The free DNA was able to enter some of the living II (R) cells and transfer to them certain III (S) characteristics. Since DNA duplicates itself during cell division, more and more cells with III (S) traits gradually developed, until there were enough to cause the pneumonia and kill the host.

This seems to be a clear-cut case of genetic transfer from the dead to the living. DNA from dead cells enters the living cells and transforms them; alters their hereditary traits. This certainly seems to imply that the genetic material is related to DNA. But even more pointed evidence will be presented in the next chapter.

CHAPTER 39

SULFUR, PHOSPHORUS AND PROTEIN GHOSTS

AS LATE AS 1950 geneticists were still searching hopefully for "the genetic material." They knew that in 1928 Griffith discovered a "transforming principle" which changed hereditary traits in pneumococci. They knew that in 1944 Avery and his group showed a close relationship between Griffith's "transforming principle" and DNA. But they remained unconvinced. No one dared go out on the limb to state flatly that DNA carries the heredity.

But in 1952, Alfred D. Hershey and Martha Chase of the Carnegie Institution performed some unusual experiments with viruses which practically forced geneticists to conclude that *DNA is the genetic material.* Before we can understand the Hershey-Chase experiments, certain facts must be made clear.

Tracer elements are radioactive isotopes which give off a steady stream of particles or rays. If atoms of a radioactive isotope are incorporated into a compound, the substance is said to be "tagged." A research scientist with the proper instruments can trace the tagged compound as it moves about in the body of an organism, or as it passes from one generation to the next.

The T_2 virus is a parasite which infects *E. coli* cells. In structure, this virus resembles a tiny medicine dropper with a bulb-shaped head, and a delicate tube-like tail. The virus backs up to an *E. coli* cell, attaches itself by means of its tail, and dissolves the bacterial wall at the point of contact. Next the bulb-shaped head contracts and injects the contents of

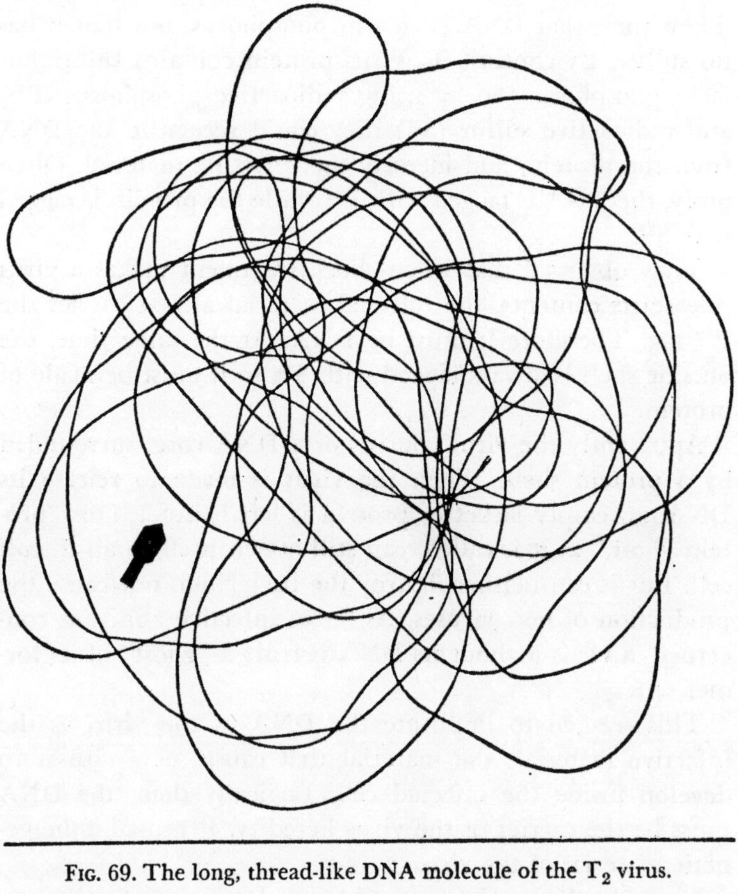

FIG. 69. The long, thread-like DNA molecule of the T_2 virus.

the virus into the bacterium. The empty virus shell remains attached to the outside of the bacterial wall.

In due course, virus reproduction occurs. The genetic material of the virus duplicates itself inside the bacterial cell. The new viruses that develop have the same hereditary material as the original infecting virus. But what was the hereditary material that infected the bacterium? Was it virus protein or virus DNA?

Hershey and Chase devised a clever way of finding out. They knew that DNA is rich in phosphorus, but that it has no sulfur. By contrast T_2 virus protein contains sulfur but lacks phosphorus. So by using radioactive phosphorus (P^{32}) and radioactive sulfur (S^{35}) they could recognize the DNA from the protein, and identify the infecting material. Obviously, the DNA is tagged with P^{32} while the protein is tagged with S^{35}.

They observed that when shock treatment makes a virus release its contents, the released material always carries the P^{32} tag. Therefore it must be DNA. At the same time, the outside shell is always tagged with S^{35}, so it must be made of protein.

Apparently the virus consists of a DNA core, surrounded by a protein shell. When the virus is made to release its DNA, an empty jacket of protein is left behind. This "protein ghost," as it is called, can still attach itself to an *E. coli* cell. But it can neither destroy the bacterium, nor cause the production of new viruses. As far as infective ability is concerned, a virus without its DNA is truly a "ghost" of its former self.

This seemed to implicate the DNA of the virus as the infective material; the material that causes new viruses to develop inside the infected cell. Logically then, the DNA must be the carrier of the virus heredity. It must be the genetic material of the virus.

Now Hershey and Chase performed a series of other experiments with T_2 viruses. In some cases only the DNA was tagged with P^{32}. In others only the protein "overcoat" was tagged with S^{35}. In still other cases, the virus was tagged with both P^{32} and S^{35}. The result was always the same. When a T_2

virus infected an *E. coli* cell, the tagged DNA entered the bacterial cell, while the tagged protein stayed outside.

They even went one step further. They tagged separate cultures of *E. coli* with P^{32} and S^{35}, and infected each culture with ordinary *untagged* T_2 viruses. The purpose of this experiment was to discover what the newly produced viruses would carry from the first bacterial host to the next.

As one would expect, viruses developing in S^{35}-tagged bacteria incorporated some radioactive sulfur into their protein. The new viruses were tagged with S^{35} that they had picked up from their bacterial host. Would they transfer any of these radioactive atoms to their next host? *The experiment showed that they did NOT.* When these viruses infected new *E. coli* cells, the *S^{35} remained outside, in the protein shells.*

In the companion experiment, viruses developing in P^{32}-tagged bacteria incorporated some radioactive atoms of phosphorus into their DNA. These new viruses were tagged with P^{32} that they had obtained from their bacterial host. Would they transfer any of these radioactive atoms to their next host? *The experiment showed that they did.* When these viruses infected new *E. coli* cells, the P^{32} entered the cell with the infecting material.

These delicate experiments indicated a very definite conclusion. When a T_2 virus infects an *E. coli* cell, only the DNA enters the cell under attack. The protective shell of protein stays outside. After attaching the virus to the bacterium, the protein shell plays no further part in the infection. The DNA does all the rest. The DNA is the infective material.

Once inside the bacterium, the virus DNA takes control of the cell mechanism. It forces the *E. coli* cell to produce more virus DNA and virus protein. Each newly produced protein shell is wrapped around a core of virus DNA to produce a complete virus particle. Soon the bacterial cell breaks open, and the new generation of viruses is released, ready and able to infect other *E. coli* cells.

These experiments left no doubt that DNA is the genetic material of the T_2 virus. And if this is true for the virus, why not for other organisms as well? The evidence was getting too strong to be ignored. Geneticists were forced to reconsider the possibility that transmission of heredity was associated with the DNA rather than with the nucleoproteins.

CHAPTER 40

TRANSDUCTION: HEREDITY BY MESSENGER

IT IS HARD to believe that a parasite can "steal" a gene from one cell and "implant" it into its next host. Yet, this is precisely what happened in a case discovered in 1952 by Norton Zinder, while working in Joshua Lederberg's laboratory. Such transfer of heredity by a messenger is called *TRANSDUCTION*. The cells involved were bacteria of the type *Salmonella typhimurium*. The transfer agent was a virus. To understand transduction one must be familiar with the events that occur when this type of virus attacks a bacterium.

1. The virus DNA enters the bacterial cell and disappears into the bacterial DNA.
2. The virus remains hidden in the bacterial cell, apparently doing nothing. It is said to be a "latent" or "temperate" virus.
3. The bacterial cell continues to live and reproduce as though nothing has happened.
4. Each time the bacterium reproduces, its DNA duplicates itself to provide each daughter cell with a full complement of genes. The virus DNA is included in the duplication.
5. Bacterial reproduction may continue for many generations while the virus remains inactive. The bacterium carries on all of its normal activities.
6. Suddenly something triggers the virus.
7. The activated virus takes control of the cell machinery,

and "orders" it to manufacture new virus DNA and new virus protein coats.

8. The virus DNA and protein coats are assembled into complete virus particles.
9. Finally, the host cell and its own DNA are completely destroyed. As the bacterial cell bursts, the new viruses are released, ready to infect other bacteria.

So far there has been no mention of genes transferred by the virus. Ordinarily the host cell and its DNA are completely destroyed. But in occasional exceptions, a bit of bacterial DNA is salvaged as described below.

A virus particle is built up when a protein "overcoat" is wrapped about a core of virus DNA. This build up occurs at the same moment that the DNA of the bacterium is breaking down. It is easy to imagine a speck of bacterial chromosome (perhaps a gene or two) lying close to a core of virus DNA. As the protein overcoat encloses the DNA core, it accidentally includes the fragment of bacterial chromosome. Now the complete virus carries its own genes plus one or two bacterial genes that don't really belong.

When this virus infects a new host, it follows the usual pattern. Once again its DNA disappears into the bacterial cell. But this time it has brought along an extra gift in the form of a couple of *bacterial* genes. The foreign bacterial genes join those already in the cell, and make themselves quite at home. Perhaps an example can serve to explain how these messenger-delivered genes may make a vital change in the activities of the infected bacterium. Consider two strains of bacteria, T and T-less.

Bacteria of strain T have a gene for synthesis of thiamine. They can live and reproduce in a culture medium lacking this vitamin because they can manufacture their own. By contrast, bacteria of strain T-less lack the gene, and cannot synthesize thiamine. They can only live on a medium that provides the vitamin.

Assume, now, that strain T is infected with a latent virus. In due course, the virus takes over and destroys the bacteria. But in the process *a few viruses pick up the bacterial gene T* for thiamine synthesis. The new crop of viruses is ready to infect other bacterial cells.

Suppose they come in contact with the T-less strain (bacteria that cannot synthesize thiamine). If a virus which has stolen a T-gene from its previous host enters a T-less cell the cell suddenly acquires the power to make thiamine. *It has acquired a T-gene by infection.* The virus messenger has brought it a legacy from a departed cousin; a legacy in the form of a tiny gene for synthesis of thiamine. With this new power, the bacterium can suddenly grow luxuriantly on a culture medium lacking in thiamine; a culture medium on which it failed to grow before.

The newly acquired T gene becomes part of its genetic make-up. It duplicates itself with the rest of the DNA in every bacterial generation. In effect then, the T-less strain of bacteria has received a bit of hereditary material by messenger, and has been converted to a T-strain.

CHAPTER 41

GENETIC TRANSFER IN BACTERIA: THREE METHODS

THIS SECTION will use simple diagrams to summarize the three methods of genetic transfer now known to occur in bacteria. In each case we will assume that a strain deficient in thiamine synthesis (T-less) is converted to a thiamine synthesizing strain (T) by acquiring the necessary gene. But in each case the active gene (T) is acquired in a slightly different way.

1. *CONJUGATION* (See section 36)

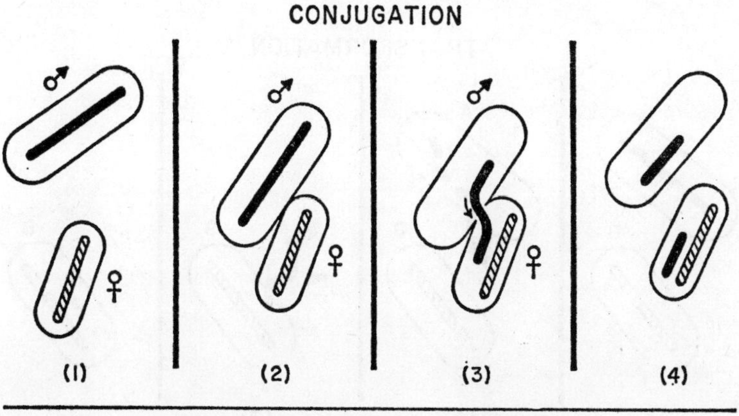

FIG. 70. Genetic transfer by *conjugation*.

The dark chromosome of the male cell carries a gene for thiamine synthesis. The striped chromosome of the female cell carries the inactive gene, and therefore the female cell cannot synthesize thiamine. The male cell comes into contact with the female cell and conjugation begins. As the male chromosome moves across the bridge, the T-gene is carried into the female cell. This cell, formerly deficient in thiamine synthesis, has now gained this ability. Even after the conjugating cells separate, the newly acquired T-gene remains part of the genetic make-up. When this cell reproduces, it gives rise to many new cells with the ability to synthesize thiamine.

2. *TRANSFORMATION* (See section 38.)

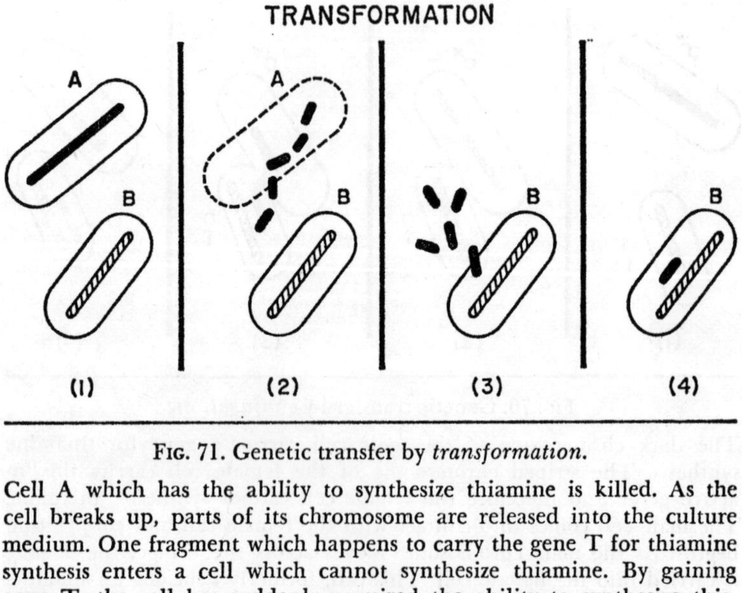

FIG. 71. Genetic transfer by *transformation.*

Cell A which has the ability to synthesize thiamine is killed. As the cell breaks up, parts of its chromosome are released into the culture medium. One fragment which happens to carry the gene T for thiamine synthesis enters a cell which cannot synthesize thiamine. By gaining gene T, the cell has suddenly acquired the ability to synthesize thiamine. When it reproduces, it passes its new-found ability on to its descendants.

3. *TRANSDUCTION* (See section 40.)

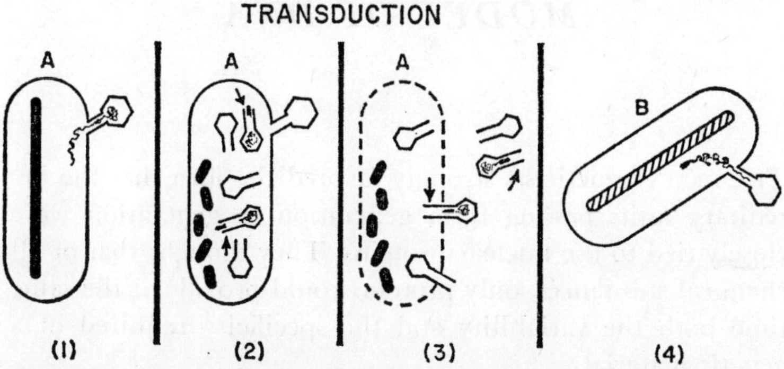

FIG. 72. Genetic transfer by *transduction.*

Cell A, with the gene for thiamine synthesis, is infected with a virus. When virus reproduction occurs, cell A and its chromosome are broken down. However, one of the new viruses (arrow) happens to pick up a fragment of bacterial chromosome carrying gene T.

This particular virus infects cell B which lacks the ability to synthesize thiamine. Together with its own DNA, the virus injects the foreign T-gene that it is carrying. The T-less cell has now acquired the ability to synthesize thiamine, an ability which it did not have before. If the virus is latent, it does not kill the cell immediately. The cell can go through many bacterial generations and produce many new cells carrying gene T and the ability to synthesize thiamine.

CHAPTER 42

THE WATSON-CRICK MODEL OF DNA

THE EARLY geneticists strongly favored the idea that the hereditary units passing from generation to generation were closely tied to the nuclear proteins. They thought that of all chemical substances, only proteins could provide at the same time both the variability and the specificity required of a genetic material.

But by the early 1950's, they were beginning to have second thoughts. The evidence seemed to favor DNA rather than proteins. Still, two problems required solving before DNA could gain final and complete acceptance as the genetic material.

1. What is the exact structure of the DNA molecule?
2. How can the structure of the DNA molecule account for the unusual characteristics basic to genetic material, namely—*variation, replication, mutation,* and *synthesis* (see section 37).

Chemical analysis of DNA had been carried out long before, so that its basic components were well known. The following description of the chemical structure of DNA was written by J. D. Watson and F. H. C. Crick. As you read it

"The chemical formula of deoxyribonucleic acid is now well established. The molecule is a very long chain, the backbone of which consists of a regular alternation of sugar and phosphate groups. . . . To each sugar is attached a nitrogenous base which can be of four different types. . . . Two of the possible bases (adenine and guanine) are purines, and the other two (thymine and cytosine) are pyrimidines."
(Watson and Crick—1953)

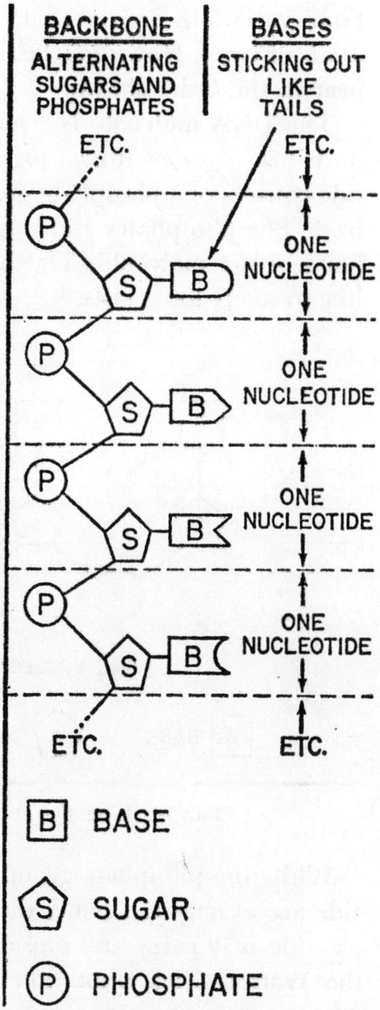

FIG. 73. The DNA molecule is a polymer consisting of a series of nucleotides.

(Fig. 73), compare it with the simplified diagram that has been placed alongside.

In the diagram P represents a phosphate group, S represents a five-carbon sugar, and B represents a nitrogenous base.

Four variations in rectangle shape represent the different kinds of bases. However, the four bases do not necessarily appear in the order shown.

The DNA molecule is a *polymer* formed of many similar units (*nucleotides*) linked together into a chain. Each nucleotide consists of a phosphate group + a sugar + a nitrogenous base. The phosphates and sugars alternate to form the backbone of the molecule. The bases stick out from the sugars like so many loose tails.

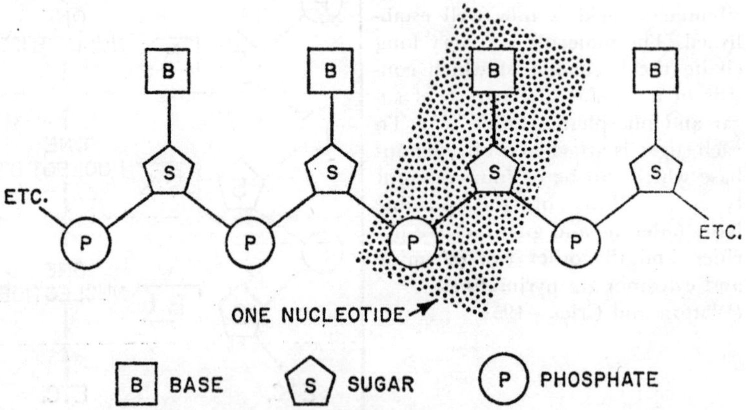

FIG. 74. A small section of a DNA molecule.

While the phosphate group and the sugar of every nucleotide are exactly the same, the bases differ. A particular nucleotide may carry any one of the four bases mentioned in the Watson-Crick quotation above (adenine, guanine, thymine, cytosine).

So, nucleic acid chemistry was well understood in 1950 when geneticists woke up to the importance of DNA. Now they had to learn more about its physical structure, and the arrangement of its units in space.

Using X-ray diffraction to analyze DNA, Dr. M. H. F.

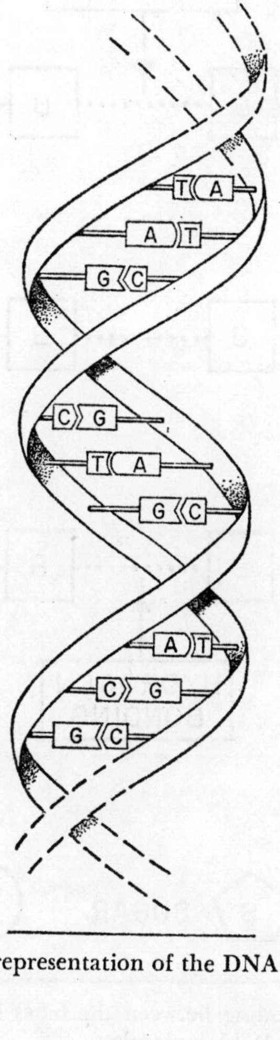

ɪɢ. 75. Diagrammatic representation of the DNA molecule as conceived by Watson and Crick.

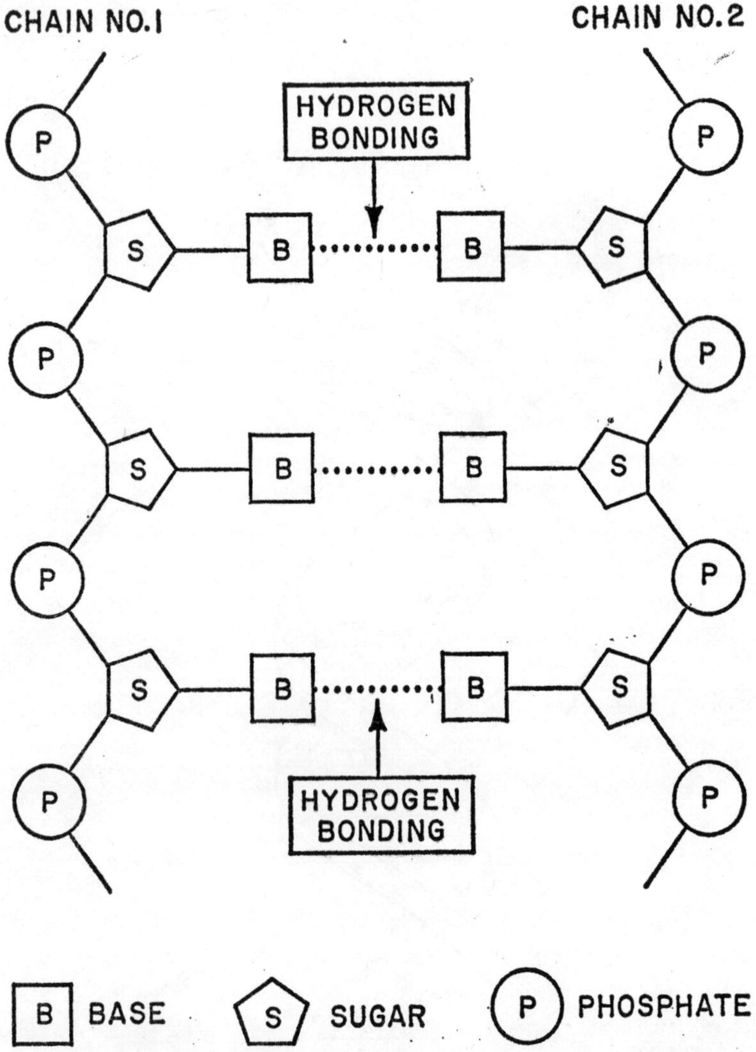

FIG. 76. Hydrogen bonding between the bases holds together the two chains making up the DNA molecule.

Wilkins produced characteristic patterns which gave information about the arrangement of atoms in the DNA molecule. He made certain measurements on these patterns and estimated sizes, distances, spaces in the molecule. The arrangement of spots on the patterns indicated to Dr. Wilkins that the DNA molecule must have a spiral shape, rather than the flat two dimensional arrangement shown in our diagram.

Many scientists now tried to build three-dimensional models based on the Wilkins data. The model suggested in 1953 by the team of Watson and Crick seems to fit the facts most closely.

The Watson-Crick model shows a double spiral formed of intertwining DNA chains, like two parallel ribbons twisted about an imaginary cylinder. The two spiral "ribbons" are really the backbones of the two DNA chains, composed of alternating sugar and phosphate groups. The nitrogenous bases attached to the sugars are arranged at right angles to the long axis of the spiral. Weak chemical forces (hydrogen bonds) between matching bases of the two intertwined chains hold the molecule together.

An analogy between the DNA molecule and a ladder may serve to clarify this picture. The uprights of the ladder correspond to the backbones of the two chains. Each rung of the ladder represents a base from one chain bonded to a complementary base from the second chain. The bonded bases (rungs) hold the uprights together. To get the three-dimensional effect, imagine that the two uprights are twisted into intertwining spirals, with the rungs still holding them together.

Watson and Crick were careful to point out a very important feature of their model. While the spirals are held together by bonding between the bases, *only certain bases can bond with one another*. Because of size and chemical structure, only complementary bases can fit together. It is like a

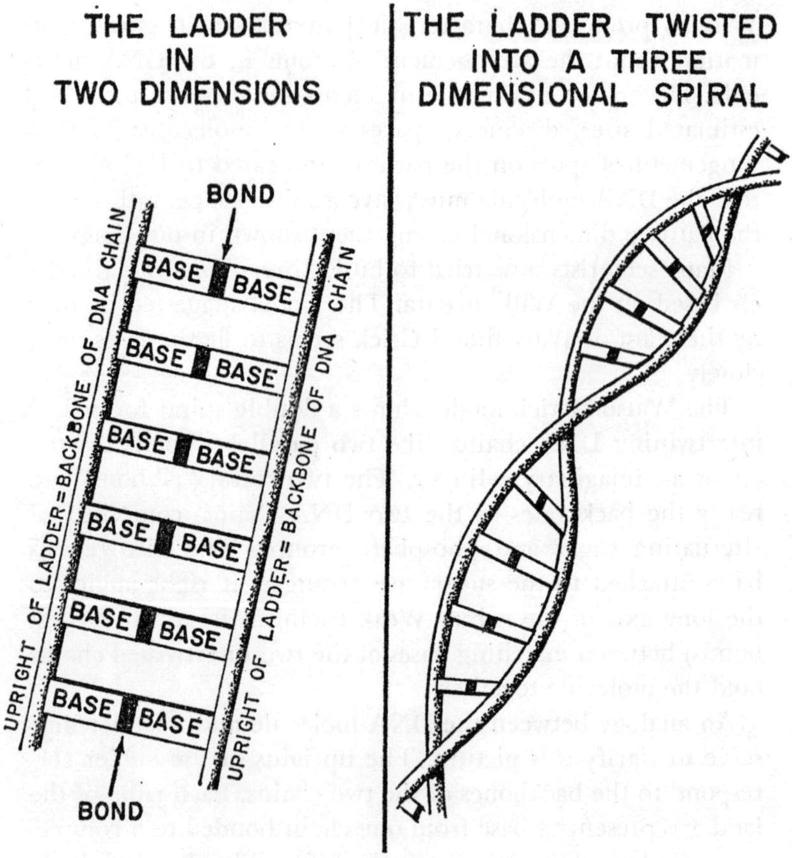

FIG. 77. The DNA molecule is like a ladder twisted into a spiral.

jig-saw puzzle, where only certain pieces can fit together. Thus, an *adenine* (A) on one chain can bond only with a *thymine* (T) on the other. A *guanine* (G) can bond only with a *cytosine* (C). Therefore, only the combinations A:T, T:A, C:G, G:C can form "rungs of the ladder." Just as the wrong pieces of a jig-saw puzzle simply don't match, so *adenine* (A) or *thymine* (T) can never bind to *cytosine* (C) or *guanine* (G).

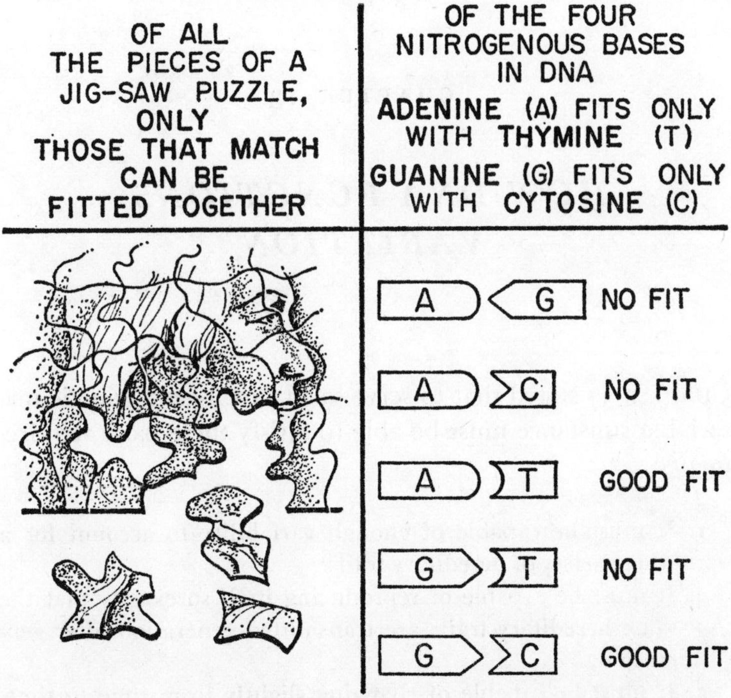

FIG. 78. The bases in the DNA molecule can fit together only in special combinations.

The Watson-Crick model of DNA seems to be satisfactory in every way. The facts and figures produced by research and testing lend it support. The model has survived since 1953 without any fundamental revision, and in 1962 Wilkins, Watson and Crick were awarded the Nobel Prize for their work in unravelling the structure of DNA.

But most important of all, the proposed model seems to be able to account for all the unusual characteristics of genetic material (see chap. 37). These characteristics will be reviewed in the next chapter.

CHAPTER 43

HOW DNA FUNCTIONS: VARIATION

CHAPTER 37 stated that to serve satisfactorily as a genetic material, a substance must be able to satisfy these basic requirements:

1. It must be capable of enough variability to account for a wide variety of hereditary traits.
2. It must be capable of reproducing itself so exactly that the same hereditary traits are transmitted generation after generation.
3. It must be capable of changing slightly from time to time, thus producing changes in hereditary traits.
4. It must be capable of synthesizing all the different kinds of molecules that participate in the development and functioning of an organism.

The evidence described in chapters 38-42 points overwhelmingly to DNA as the genetic material. It now remains for us to see whether DNA can in fact satisfy the conditions outlined above. The rest of this chapter will consider the first criterion—*variation.* The question to keep in mind is this. Does the Watson-Crick model of DNA suggest a substance capable of enough genetic variation to account for all the different hereditary traits transmitted by the many organisms of the world?

At first glance (chap. 42) the structure of DNA seems too

rigid and inflexible to allow for wide variability. The sugar and phosphate groups that make up the intertwining chains *NEVER* vary. This means that any variation that does exist must be explained by the bases . . . and there are only four of these—*adenine* (A), *thymine* (T), *guanine* (G) and *cytosine* (C). How can four bases possibly produce such a wide variety of hereditary traits as we know exist?

But think of the Morse Code. Every letter of the alphabet is represented by a different combination of dots and dashes. (A = ·—, B = —···, C = —·—·, etc.) These coded letters can be arranged to form any word of the English language. (AS = ·— ···, SOB = ··· ——— —···, BOSS = —··· ——— ··· ···, etc.)

Here is a code based on *TWO* symbols only. Yet, it provides enough variability to transmit unlimited numbers of complex messages—words, sentences, paragraphs or even whole stories. The four bases in DNA may be considered as four symbols of another kind of code—a *genetic* code. These four symbols transmit the "words, sentences, paragraphs, and stories" in the language of heredity.

Obviously, a four-symbol code is capable of far more variability than a two-symbol code. And if a two-symbol code can express the countless variations in words and ideas of the English language, then a four-symbol code should be able to express the even greater variation required by the genetic language.

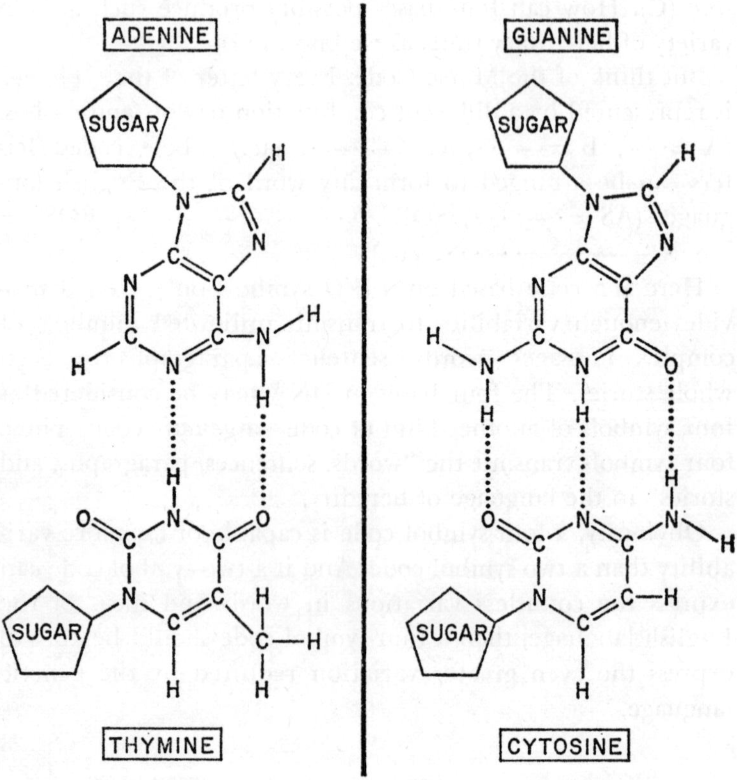

FIG. 79. Hydrogen bonding between the base pairs of DNA.

CHAPTER 44

HOW DNA FUNCTIONS · REPLICATION

IN DESCRIBING their model of DNA, Watson and Crick made the following statement:

"If the actual order of the bases on one chain of a pair were given, one could write down the exact order of the bases on the other one because of the specific pairing. Thus one chain . . . is the complement of the other. It is this feature which suggests how the deoxyribonucleic acid might duplicate itself."

This statement refers to the fact that the chemical structure of *adenine* (A) allows it to bond only with *thymine* (T); while *cytosine* (C) can bond only with *guanine* (G). Thus, DNA contains two complementary pairs of bases: adenine–thymine (A:T or T:A) and cystosine–guanine (C:G or G:C).

Therefore the sequence on the second chain is fixed by the sequence of bases on the first chain. Note that wherever one chain carries a thymine (T), the other *MUST* carry an adenine (A). Wherever there is a cytosine (C) on one chain, there *MUST* be a guanine (G) on the other.

If one chain carries
this sequence → ATTCTGAGGTCCAAAG. . . . . . etc
The other chain
MUST be → TAAGACTCCAGGTTTC. . . . . etc

However, during nuclear division, something weakens the hydrogen bonds between A:T and C:G. As these bonds gradually let go, the DNA molecule "unzips" into two separate single strands.

This unzipped condition does not last for long. The cell contains a random collection of nucleotides, carrying the bases *adenine, thymine, cytosine* or *guanine*. These nucleotides are activated and ready to bond with complementary bases.

Even as the double chain is "unzipping," the free, active nucleotides seek out spots where they fit. Every free T-nucleotide lines up opposite an A-nucleotide in the single chain; free G-nucleotides opposite C-nucleotides; free C-nucleotides opposite G-nucleotides; and free A-nucleotides opposite T-nucleotides.

The result is shown in the diagram. As the original double-stranded DNA "unzips," each single strand accumulates onto itself a series of complementary nucleotides. Single strands become double-stranded again as new chains build up along side. The new chains are the exact complements of the originals.

When the process is finished, there must be two identical double-stranded molecules of DNA, with exactly the same sequence of bases. They are as identical as identical twins can be. *The DNA molecule is replicated exactly.*

This theoretical description is strongly supported by experimental work. The 1959 Nobel Prize was awarded to Dr. Arthur Kornberg for his experimental demonstration that DNA can replicate itself in a test tube. He mixed together the basic ingredients of which DNA is made—some five-carbon sugars, some tri-phosphates, and equal quantities of activated nucleotides carrying the four bases. To this he added an enzyme extracted from the bacterium *E. coli*. And finally he added a small quantity of existing DNA.

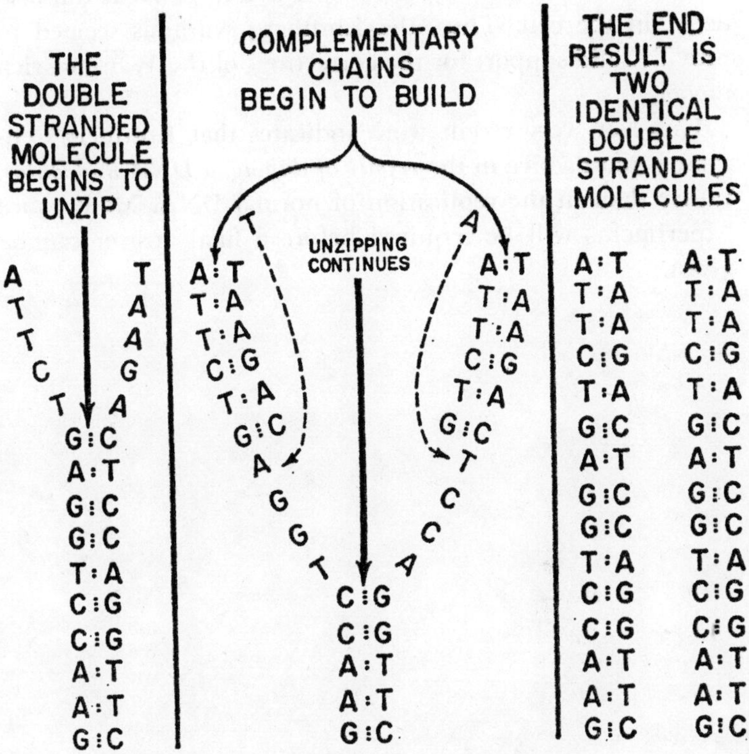

FIG. 80. How the DNA molecule duplicates itself exactly.

After incubation, Dr. Kornberg was able to recover from his test tubes more DNA than he had originally added. But most important: *the new DNA was identical with the original* bit that had been put in to prime the reaction.

Evidently the DNA had replicated itself exactly. If replication can occur in the test tube, it is likely that it can also occur in the cell. Thus, the Kornberg synthesis seemed to provide solid support for the correctness of the Watson-Crick model.

However, very recent work indicates that Kornberg's enzyme is most active in the *repair of damaged DNA molecules,* rather than in the replication of normal DNA. Many more experiments will be required before a final answer can be given.

CHAPTER 45

HOW DNA FUNCTIONS: MUTATION

Most geneticists today accept the idea that a given sequence of bases in the DNA molecule carries a coded message concerning heredity. Because the DNA molecule duplicates itself so faithfully, the precise sequence of bases is always preserved. Therefore the message is usually transmitted unchanged for many generations.

But if we accept this idea we must agree that any accidental change in sequence can alter the message slightly. The message is still transmitted, but it is somewhat different, with altered instructions. In obeying these new instructions, the developing cells do something different, and perhaps the result is an altered trait. Such a changed trait is called a mutation. It can be inherited just like the original trait, because the new sequence of nucleotides is duplicated just as faithfully as the original sequence.

How large does the change have to be to alter the message? We really cannot give an exact answer to this question. But a simple analogy may provide a clue. A sentence can be considered to be a coded message carried by a certain sequence of letters. The brain decodes the message and registers some sort of interpretation.

With this in mind, consider the effect of a simple typing error. The copy reads: HE WAS A CHIEF. The typist prints: HE WAS A THIEF. *The two coded messages differ by a single letter.* But what different pictures they conjure up in the reader's mind! Just one simple typographical error, the substitution of a *T* for a *C,* was enough to have somebody sued for libel.

We can conceive that the same thing happens in a DNA molecule. We can visualize a change in a single base of a nucleotide sequence. But it is hard to visualize the final effect of such a minute change. It is hard to say that a change in a single base will cause a visible mutation.

The case of sickle-cell anemia (section 27) brings us about as close as we have come. Ingram showed that a single amino acid on a chain of 300 differentiated normal hemoglobin-A from sickle-cell hemoglobin-S. We say that this change came about when gene-A mutated to gene-S. Evidently there was some sort of typographical error in the coded message that instructed the body to produce a new kind of hemoglobin. But how many DNA molecules were involved in this mutation? How many bases were altered? This we do not know.

There is no reason to believe that a mutation involves a change in only one nucleotide. Probably many nucleotides must be altered before a recognizable change in trait is produced; before a visible mutation is developed.

On the other hand, such a change may be possible without the alteration of a single nucleotide. Perhaps there need only be a reversal of part of the sequence. Thus, the sentence HE *WAS* A THIEF means something quite different from the sentence HE *SAW* A THIEF. Yet both sentences contain exactly the same letters; exactly the same code symbols. The reversal in sequence of only three of the letters made the message come out quite different.

Consider another analogy involving the letters O-P-S-T. Arranged *POTS* they carry a coded message which the brain interprets as "cooking utensils." But a slight rearrangement makes STOP or TOPS or POST or SPOT. Each slight variation in sequence carries a completely different coded message.

Can the same be said of DNA? If the sequence ATCG means something in the language of heredity, is it possible that a slight rearrangement to GATC or GTCA or AGTC means something different? We do not really know, but we must assume that such a rearrangement in sequence might spell out a change in trait.

However, we do know that nucleotides can change; that they are changed by the same conditions which cause mutations. Agents like X-rays, neutrons, atomic radiation, mustard gas, etc. are most apt to affect DNA.

We also know that DNA molecules with changed nucleotides reproduce themselves in their new form just as faithfully as the original unchanged molecules used to reproduce themselves. Taken all together, these facts tend to indicate that DNA can indeed account for the changes in living things that we call mutations.

This is a good place to mention a completely unexpected discovery. Certain delicate experiments indicate that DNA molecules damaged by radiation can repair themselves. The damaged portion is cut out and destroyed. Then the DNA molecule is rebuilt in its perfectly normal state. If these experiments are actually correct, we must conclude that living organisms have an amazing built-in mechanism to protect their genetic inheritance. Perhaps this can help explain why hereditary traits remain so stable despite severe environmental changes.

CHAPTER 46

HOW DNA FUNCTIONS: SYNTHESIS

CHAPTERS 42-45 explained how the structure of DNA can account for genetic variation, mutation and exact replication. Another important characteristic remains to be considered, as pointed out in the following quotation from Dr. Arthur Kornberg's 1959 Nobel Prize lecture. *"Deoxyribonucleic acid, like a tape recording, carries a message in which there are specific instructions for a job to be done."* Dr. Kornberg is referring to the job of synthesis.

A living cell is a complex chemical factory which can synthesize a wide variety of proteins, enzymes, vitamins, hormones, toxins, antibodies, tissue components, etc. We believe that the instructions for each specific synthesis are carried in coded form in the DNA molecules of the nucleus.

But let us forget cells for a moment to consider a huge, automated factory that produces 1000 different items for the market. Naturally, the company designers have prepared plans or blue-prints for every one of the items. Each plan tells exactly what raw materials are needed, and how these raw materials must be put together to make the finished product. Since this is an automated factory, we will assume that instructions for each item have been punched out on special cards. These cards are carefully classified and arranged in a special file room in the front office, where the manager can locate any one at a moment's notice.

Another part of the factory contains the assembly line of machines that do the actual work. These same machines can produce any one of the 1000 products, but they must be "told" which one to make. This is done by getting the proper punched card from the files and inserting it into the control box. If card #364 is inserted, the machine begins to assemble certain raw materials in the particular way that produces item #364. But if card #734 is put into the control box, a different sequence of operations is set in motion, different raw materials are used, and a different product results.

Of course, raw materials are needed to make every product. Therefore the factory must have a stockroom, with a supply of every raw material that may be required on the assembly line. It is likely that the same raw materials are used to make many different products. If this particular factory uses just 20 basic raw materials in producing all of its 1000 products, then the stockroom must have 20 different bins, each containing a supply of its own particular raw material.

There must also be some way of coordinating the whole operation; some method of communication between front office, stockroom, and assembly line. Thus, if the front office receives an order for item #822, the manager goes to the file and lifts out the punched card of instructions. This card is delivered to the assembly line to activate the machinery, and make it operate in the correct way to produce item #822. But at the same time, the stock-clerks must be notified to deliver certain quantities of particular raw materials from the stock-room to the assembly line. If all this is properly timed, the machines begin turning out item #822.

We think that synthesis in a cell "factory" operates in just about the same way. The nucleus is the "front office." The chromosomes are the "filing cabinets" containing DNA "blueprints" for many different kinds of synthesis. Raw materials like amino acids are usually obtained from food and stockpiled in the cytoplasm of the cell. The "assembly line" is composed of a host of tiny structures in the cytoplasm called *RIBOSOMES*. These seem to be the centers for protein synthesis.

In this analogy, the DNA is compared to the blue-print for a construction job, which tells what materials to use and how to assemble them into a complete structure. But the blue-print itself cannot do the work. Someone or something must read the instructions and perform the operations as directed. So the question comes down to this:

How do the instructions coded in the DNA "blue-prints" reach the working parts of the cell, and how are they interpreted once they get there?

This question will be considered in the next chapter.

CHAPTER 47

HOW DNA FUNCTIONS: RELATION TO RNA

THE QUESTIONS raised at the end of the last chapter can be restated more specifically as follows: How does the information stored in the DNA of the nucleus reach the ribosomes to tell them what proteins to make. And when instructed to make a specific substance like myosin or pepsin or insulin, how do the ribosomes assemble the necessary amino acids in the correct order to produce the long, complex protein chain?

To help answer these questions we must introduce another substance; a chemical "cousin" to DNA, called *RIBONUCLEIC ACID* or *RNA*. Ribonucleic acid (RNA) and deoxyribonucleic acid (DNA) are both chain-like molecules built of nucleotides. But there are several differences.

1. The five-carbon ribose sugar of RNA contains one more oxygen than the five-carbon deoxyribose sugar of DNA.
2. The bases in DNA are guanine (G), cytosine (C), adenine (A), and thymine (T), while the bases of RNA are guanine, cytosine, adenine and uracil (U). The base thymine of DNA is replaced by uracil, a closely related compound.
3. RNA may consist of a single spiral strand rather than the usual double strand of DNA.
4. RNA molecules may be comparatively short, with only a few nucleotides, while DNA is a relatively long chain of nucleotides.

5. While DNA is usually found only in the nucleus of a cell, RNA may be found in the nucleus, in the cytoplasm, and in the ribosomes.

The presence of RNA in both nucleus and cytoplasm suggests an interesting theory. Perhaps messages are transferred from DNA in the chromosomes to smaller RNA molecules which leave the nucleus and migrate into the cytoplasm. Perhaps RNA molecules are really "messenger-boys" carrying instructions from the DNA "blue-prints" to the cell machinery. The name *MESSENGER RNA* suggests itself.

But the ribosomes (where protein synthesis occurs) are also rich in RNA. This kind of RNA does not seem to migrate. It is called *RIBOSOMAL RNA* because it remains fixed in the ribosomes.

Some very short-chain RNA molecules were also discovered in the cytoplasm. These can pick up amino acid molecules and transfer them to other places in the cell. The name *TRANSFER RNA* seems quite appropriate for this kind of RNA.

These three distinct types of RNA help to coordinate the manufacturing operations of the cell. The following outline describes the process in greatly *oversimplified* form.

1. The nucleus directs all the manufacturing processes of the cell. It is the "front office" so to speak. The chromosomes are living "file-cabinets" containing many DNA "blue-prints." Each blue-print gives coded instructions for the synthesis of a specific substance.
2. When the chromosomes are actively directing cell activities, certain DNA "blue-prints" are "pulled out of the file" and copied. It is important to note that not all DNA molecules are equally active at all times.
3. During the copying process, active DNA molecules serve as molds against which RNA MESSENGER molecules are built up. Thus, an exact copy of the information coded in

the DNA "blue-print" is transferred to the MESSENGER RNA.

4. The MESSENGER RNA separates away from the DNA model which it copied. It migrates out of the nucleus and attaches itself to a group of ribosomes somewhere in the cell.
5. The MESSENGER RNA associated with RIBOSOMAL RNA is ready to serve as a plan or model (template) for protein synthesis. It "tells" the machinery exactly how to organize the raw materials into the finished product.
6. In this case, the raw materials are amino acids which must be hooked together in the proper order to create a particular protein. Living things build all their proteins from about 20 different amino acids. These are usually obtained from the digested food, and stockpiled in the cytoplasm.
7. Amino acids must be activated before they can be hooked together into the long protein chains. This happens in the cytoplasm when the amino acid reacts with energy-rich ATP under the influence of a special enzyme.
8. An activated amino acid can be picked up by a molecule of TRANSFER RNA, and attached like a tail to one end of the RNA molecule. There is a specific kind of TRANSFER RNA for each kind of amino acid. Thus, the amino acid *glycine* is picked up by a *GLYCINE*-TRANSFER-RNA, *alanine* is picked up by an *ALANINE*-TRANSFER-RNA, etc. There must be at least 20 different kinds of TRANSFER RNA'S to transport the 20 different kinds of amino acids.
9. The TRANSFER RNA's act like vehicles that carry activated amino acids to places where they are needed. Each TRANSFER RNA has its own particular "chemical shape" and carries its own brand of amino acid.
10. All kinds of TRANSFER RNA's approach the ribosomes to which a molecule of MESSENGER RNA is fixed. They are attracted to empty sites on the MESSENGER RNA, but they can only settle into those sites which have the proper "chemical shape" (just as a piece of a jig-saw puzzle can only fit together with another piece of the right shape).

11. The different TRANSFER RNA's that line up along the MESSENGER RNA chain *must* be in the order dictated by the instructions on the MESSENGER RNA. They cannot fit any other way. This automatically places the amino acids in correct sequence for the particular protein being synthesized.
12. Once the amino acids have been placed in proper sequence, they can bond together into the long, complex protein chain. A molecule of protein has been synthesized.
13. The finished protein molecule separates away. The TRANSFER RNA's are released, and they leave their sites on the MESSENGER RNA. Each TRANSFER RNA is now free to escort another amino acid of the proper kind to another appropriate site.
14. The MESSENGER RNA remains attached to the ribosomes. Just as a rubber stamp is used over and over again to make the identical impression, so the MESSENGER RNA is used over and over again to model molecule after molecule of the same protein.

Thus, in summary: the DNA sets a basic pattern for protein synthesis, and the three kinds of RNA cooperate to carry out this pattern. The next chapter summarizes this concept, but in diagramatic form.

In the diagram it is assumed that the protein chain being synthesized consists of four units, the four amino acids *valine*, *alanine*, *phenylalanine*, and *lysine*, in that order. Of course, this is strictly an imaginary protein-like substance (a polypeptide). Furthermore, the diagrams present a theory that is still far from being fully confirmed.

CHAPTER 48

A THEORY OF PROTEIN SYNTHESIS DIAGRAMATICALLY EXPLAINED

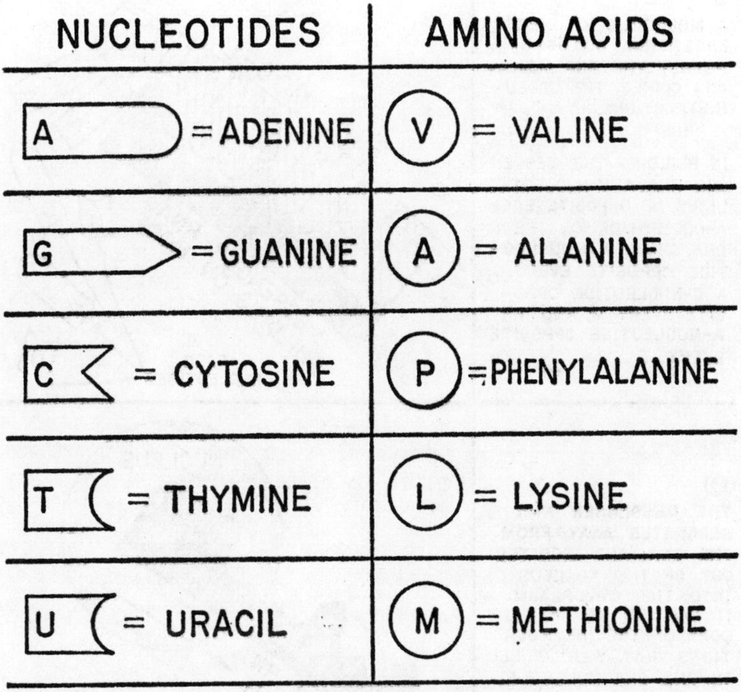

Fig. 81. A theory of protein synthesis.

This page provides a key to the symbols used in the six figures that outline a theory of protein synthesis. The six figures appear on the facing pages that follow immediately after the key.

① THIS REPRESENTS A SMALL PORTION OF A DNA CHAIN IN THE NUCLEUS, CARRYING THE "BLUEPRINT" FOR THE IMAGINARY POLYPEPTIDE V–A–P–L.

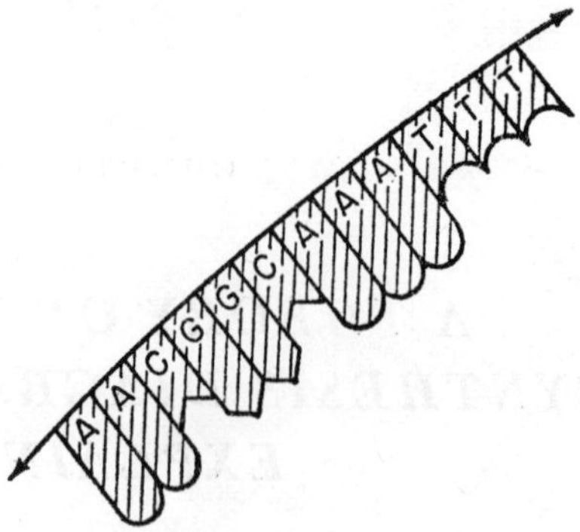

② A MOLECULE OF MESSENGER RNA BUILDS UP AGAINST THE DNA MODEL, AND COPIES THE CODED INSTRUCTIONS.

IN BUILDING THE MESSENGER RNA, A U-NUCLEOTIDE LINES UP OPPOSITE EACH A-NUCLEOTIDE ON THE DNA CHAIN; A G-NUCLEOTIDE OPPOSITE EVERY C; A C-NUCLEOTIDE OPPOSITE EVERY G; AND AN A-NUCLEOTIDE OPPOSITE EVERY T.

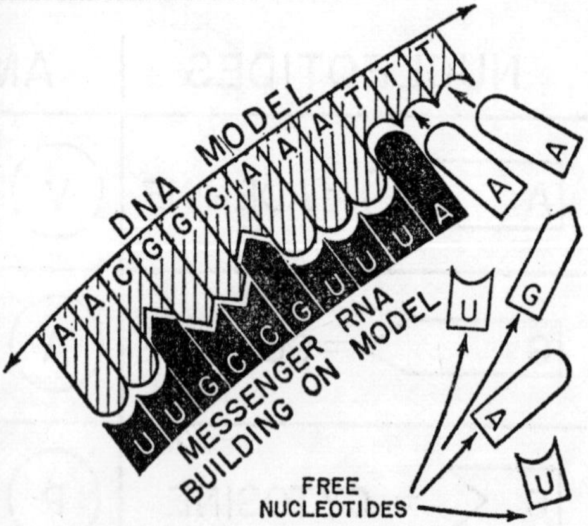

③ THE MESSENGER RNA SEPARATES AWAY FROM THE DNA, AND MIGRATES OUT OF THE NUCLEUS INTO THE CYTOPLASM. IT CARRIES AN EXACT COPY OF THE INSTRUCTIONS THAT WERE CODED IN THE DNA MOLECULE.

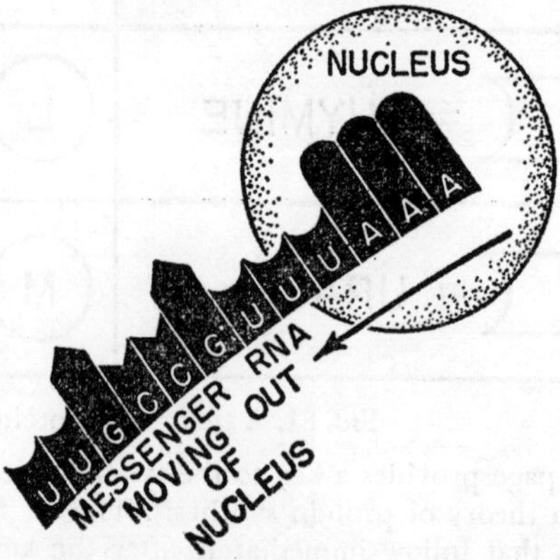

④

THE MESSENGER RNA ATTACHES ITSELF TO SEVERAL RIBOSOMES.

VARIOUS TRANSFER RNA'S APPROACH, EACH BEARING ITS OWN KIND OF ACTIVATED AMINO ACID.

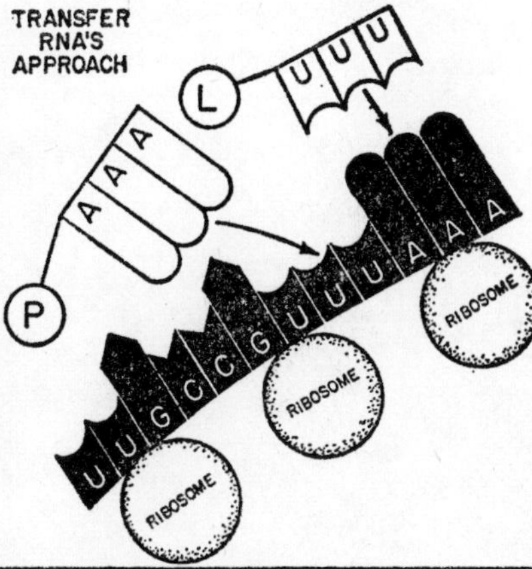

⑤

TRANSFER RNA'S THAT FIT ARE ATTRACTED TO SPECIFIC SITES ON THE MESSENGER RNA.

IN THE ILLUSTRATION, METHIONINE-TRANSFER-RNA CANNOT FIND AN APPROPRIATE SITE, WHILE THE OTHERS DO.

THE AMINO ACIDS VALINE, ALANINE, PHENYLALANINE, AND LYSINE ARE LINED UP IN THE PROPER POSITION. THEY BOND AS SHOWN TO FORM THE CHAIN V–A–P–L.

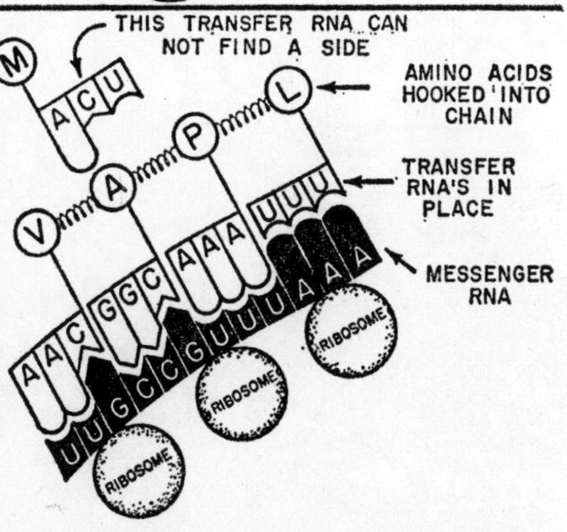

⑥

A PROTEIN-LIKE SUBSTANCE MADE OF FOUR AMINO ACIDS HAS BEEN BUILT UP. IT SEPARATES AWAY FROM THE TRANSFER RNA'S THAT BROUGHT THE INDIVIDUAL AMINO ACIDS.

THE TRANSFER RNA'S ARE FREED FROM THE MESSENGER RNA. EACH ONE CAN NOW PICK UP ANOTHER AMINO ACID OF THE RIGHT TYPE, AND TRANSFER IT TO ANOTHER SITE OF PROTEIN SYNTHESIS.

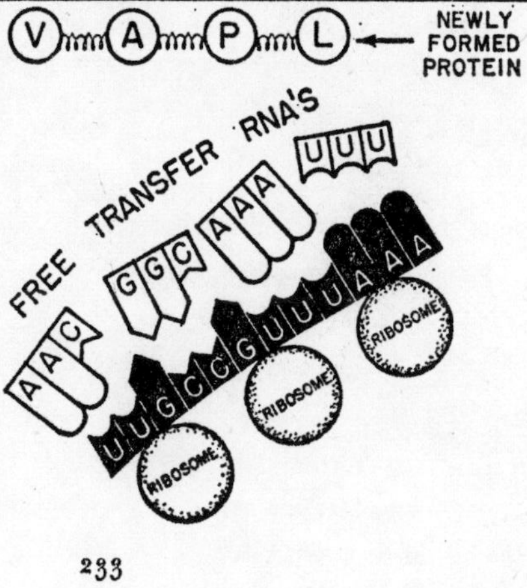

CHAPTER 49

THE GENETIC CODE

ANOTHER PROBLEM facing modern geneticists is that of *cracking the genetic code.* We cannot read the information carried by the DNA blue-print, unless we understand the code in which it is written. Without knowing the code we are in the position of a man who finds a coded message describing the location of a great treasure. He can never locate the treasure unless he first cracks the code.

Cracking any code depends on taking advantage of every possible hint. In the case of the genetic code, we have one important clue to start us off. The secret messages of heredity are hidden in the DNA molecules, and since DNA is nothing more than a long chain of nucleotides, *the nucleotides must be the code symbols.*

In terms of protein synthesis, the sequence of nucleotides must indicate the sequence of amino acids in the protein molecule. Therefore the problem can be restated as follows: *What nucleotide or sequence of nucleotides indicates each particular amino acid?*

We know that twenty different amino acids are used in synthesizing the proteins of living things. Therefore, there must be twenty different symbols in the code—one for each particular amino acid. But DNA carries only four different nucleotides: adenine (A), thymine (T), guanine (G), and cytosine (C). If each of these symobls were to represent a single amino acid, then the DNA could specify only four amino acids, as explained in the next paragraph.

Let us assume that nucleotide A = the amino acid proline, T = glycine, G = tyrosine and C = arginine. The entire genetic alphabet has been used up to specify only four distinct amino acids. Sixteen other amino acids are left over without symbols. Obviously, a "single-symbol code" in which one nucleotide equals one amino acid is quite impossible.

On the other hand, it is just as impossible to have the same symbol represent several different amino acids. If nucleotide A meant proline at one time, glycine at another time, threonine at still another time, the blue-print could never be read with any degree of accuracy. And we know that a protein is always built with such exactness that the amino acid chain is precisely the same each time the specific protein is synthesized.

The only logical conclusion is that each given amino acid is represented by a group of nucleotides, rather than by a single nucleotide.

Using nucleotides in pairs, like AA, AT, AC, CG, etc. gives only sixteen possible combinations ($4^2 = 16$). Sixteen symbols cannot account for twenty amino acids. However, using nucleotides in triplets, like ACG, AAA, GTC, etc. allows for 64 combinations ($4^3 = 64$). This provides more than enough symbols to code twenty amino acids.

Logic like this provided one clue for the cracking of the genetic code. It was reasonable to assume that the smallest group of nucleotides that could serve as the code symbol for one particular amino acid must be "a triplet." A certain group of three nucleotides specifies the amino acid proline, another combination of three nucleotides represents the amino acid serine, a third triplet symbolizes tyrosine, and so on. Note, that a triplet is the *smallest* group that can serve. There was nothing up to this point to rule out the possibility that a sequence of four, or five, or six nucleotides might be required to specify an amino acid.

As we know, scientists are not satisfied with just plain logic, to solve problems. They depend on experiments and on direct observation wherever this is possible. And in 1962, Drs. Nirenberg and Matthaie provided the first experimental "crack" in the code. They prepared an artificial MESSENGER RNA from a single nucleotide base—*uracil*. This MESSENGER RNA can be called POLY-U MESSENGER RNA, because it consists of many uracil nucleotides hooked together in the RNA chain. (-U-U-U-U-U-U-U-U-etc.)

Next, the POLY-U MESSENGER RNA was incubated with a mixture of ingredients obtained from the bacterium *E. coli*. The incubating system contained no whole cells, but it did contain ribosomes, various types of TRANSFER RNA's, amino acids, and enzymes. Evidently synthesis went on, even in the absence of whole cells, because amino acids were linked up to form short, protein-like chains.

However, these were unusual protein chains. They were built of exactly one amino acid (phenylalanine), repeating itself over and over again in the chain. The conclusion was obvious. The only kind of TRANSFER RNA that could find a suitable site on the POLY-U MESSENGER RNA was PHENYLALANINE TRANSFER RNA. Other types of TRANSFER RNA's did not fit, so their amino acids could not possibly participate in this particular synthesis.

This experiment indicated clearly that the MESSENGER RNA code symbol for the amino acid phenylalanine was U-U-U- at least in the bacterium *E. coli*. Since the MESSENGER RNA was formed against a DNA model, the U-U-U must have developed against A-A-A on the DNA molecule. Similarly, the only kind of TRANSFER RNA that can find a suitable site on the U-U-U of the MESSENGER RNA is one that is made of three adenine nucleotides -A-A-A. Perhaps the diagram below will clarify this.

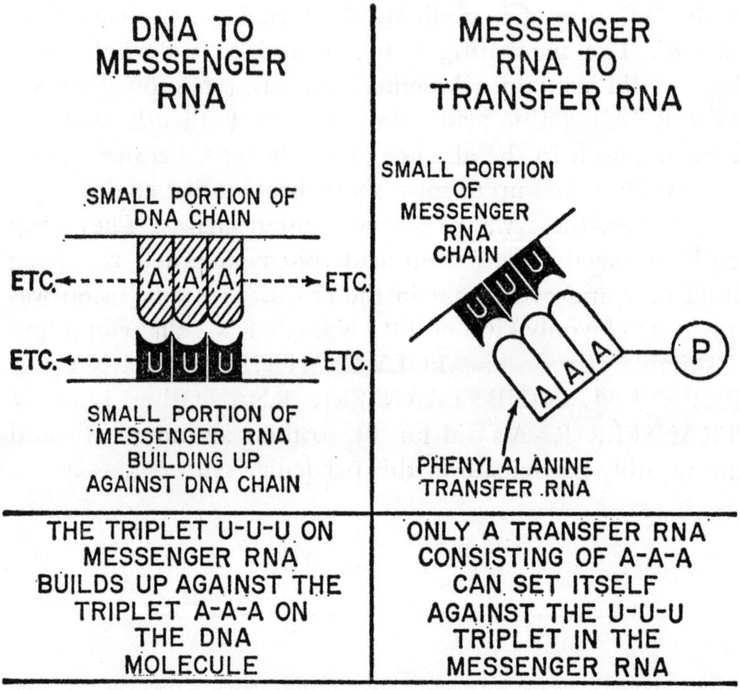

Fig. 82. The first crack in the genetic code.

Once Nirenberg and Matthaie set the pattern, other artificial MESSENGER RNA's were prepared, and used to discover what kind of proteins they would build. These experiments led to the establishment of a tentative *E. coli* code for all twenty amino acids. Further tests indicated that the *E. coli* code held good for all sorts of animals and plants. In fact, the code seemed to be quite universal.

The code that is usually discussed is the MESSENGER RNA code. Obviously, on the original DNA blue-print the code must appear as the complementary triplet (see diagram). Likewise, the TRANSFER RNA triplet must be similar to the DNA triplet, except that a uracil nucleotide appears wherever there was a thymine nucleotide on the DNA. The table below summarizes the code symbols for each of the twenty amino acids.

In most cases, the exact order of the three nucleotides making up a triplet is not yet known. Therefore, if the table says that UGA is the code symbol on MESSENGER RNA for methionine, it does not necessarily mean that the order is U-G-A. It could just as well be G-U-A, or A-G-U or some other arrangement of those three nucleotides. We are only making a start at determining the exact order.

Note also that the triplets given in the table are not necessarily the only ones capable of symbolizing the particular amino acids. In some experiments additional triplets seemed to signal the same amino acid as the triplet given here. Besides, all of the triplets given here are based on experiments in cell-free systems. We can only *assume* that the same triplets will work the same way inside the living cell.

Amino Acid	*On DNA Molecule*	*The code on Messenger RNA**	*On Transfer RNA*
1. Alanine	GGC	CCG	GGC
2. Arginine	GCG	CGC	GCG
3. Asparagine	TGT	ACA	UGU
4. Aspartic Acid	CAT	GUA	CAU
5. Cysteine	AAC	UUG	AAC
6. Glutamic Acid	CTT	GAA	CUU
7. Glutamine	TTG	AAC	UUG
8. Glycine	ACC	UGG	ACC
9. Histidine	TGG	ACC	UGG
10. Isoleucine	ATA	UAU	AUA
11. Leucine	AAC	UUG	AAC
12. Lysine	TTT	AAA	UUU
13. Methionine	ACT	UGA	ACU
14. Phenylalanine	AAA	UUU	AAA
15. Proline	GGG	CCC	GGG
16. Serine	AGA	UCU	AGA
17. Threonine	GTG	CAC	GUG
18. Tryptophan	CCA	GGU	CCA
19. Tyrosine	TAA	AUU	UAA
20. Valine	ACA	UGU	ACA

* The Code on Messenger RNA—after M. W. Nirenberg, 1963.

FIG. 83. The Genetic Code. (The triplets given are not necessarily the only ones capable of signaling a particular amino acid.)

Another interesting facet of the DNA code that still needs deciphering is the matter of punctuation. The meaning of a passage may be changed drastically by a difference in punctuation. If you are skeptical of this, try reading the passage below in the two ways indicated by punctuation marks.

Every lady in the land
has 20 fingers on each hand,
Five and twenty
    on hands and feet.
All this is true
    without deceit.

Every lady in the land
Has 20 fingers. On each hand
Five, and twenty
    on hands and feet.
All this is true
    without deceit.

In the same way, the meaning of the coded sentence on the DNA blue-print may change according to the "punctuation." What is there in DNA to show where one message ends and another begins? What is there in a MESSENGER RNA molecule to show where one triplet ends and the next one begins?

As far as the DNA is concerned, Dr. Aaron Bendich and coworkers at Sloan Kettering Institute think that the messages are punctuated by amino acids or short protein-like substances inserted into the chain at intervals. Whether or not they are correct, remains to be seen.

Many more experiments will be required before we can say that we fully understand the genetic code. But the first results have already been achieved. The code is in the process of being cracked.

CHAPTER 50

WHAT IS INHERITED—HEREDITY AND ENVIRONMENT

WHAT IS really inherited? Does a green plant inherit green color from its green parents? Does a child inherit low intelligence from its unintelligent parents? A little consideration provides the answer. The plant raised in the dark remains white, even though it has genes for green color. Did it inherit green color, or did it inherit the ability to produce green color in the presence of light? The I.Q.'s of children can often be raised merely by providing a better diet. Did these children inherit a certain degree of intelligence, or did they inherit the capacity to develop a certain degree of intelligence, given the proper conditions?

It is difficult to unravel heredity from environment. The two must work hand in hand. What is passed on from a parent to a child is a set of genes. But these genes cannot operate in a vacuum. They must express themselves in an environment of some sort, and the organism which develops is the product of the interaction between its heredity and its environment. Consider, for example, the case illustrated in Fig. 84. What did these fruit flies really inherit?

Clearly, what we inherit is not a characteristic or a trait; rather, we inherit a capacity. We inherit the ability to react to a situation. These capacities and abilities can be developed if the proper environment is provided, or they may remain repressed if the opportunity for development is not pro-

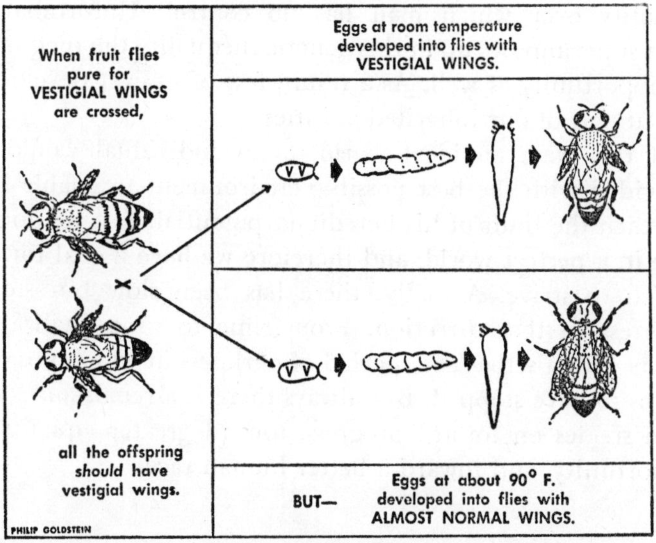

FIG. 84. What is inherited?

vided. Thus, many European children with genes for good physique will remain stunted throughout life because they were starved and maltreated during the Nazi occupation. Many Americans remain intellectually stunted because educational opportunities are lacking, or because diets are inadequate.

Fortunately, genes are highly resistant to outside forces. Though the starved children become stunted adults, they still pass on to their children the capacity to develop sound, well-built bodies. Similarly, those whose intellectual growth was stunted by lack of opportunity can still pass on to their children the capacity to develop high intelligence.

Of course, all of us do not inherit the same capacities. Genetically, all men are *not* created equal. We know that each of us is endowed with a different set of genes, and that the combination of genes in any set resulted largely from the action of chance. Thus the mechanics of heredity create inequality over which man has no control. Unfortunately, man superimposes upon this genetic inequality the inequality of opportunity as well. As a result, few of us ever develop to the utmost of our inherited abilities.

If this were an ideal world, every individual would be provided with the best possible environment to enable him to reach the limit of his hereditary potential. But we do not live in a perfect world, and therefore we have a goal toward which to strive. Actually, there has been slow but steady progress in that direction. From time to time civilization meets with a setback—a period of darkness in which progress seems to have stopped. But always there is a renaissance, and man strides on toward progress, toward greater equality of opportunity, and toward a better human race.

CHAPTER 51

IS IT HEREDITARY?

HOW CAN you tell if a trait is hereditary?

Fig. 85 illustrates the pedigree of a family in which pellagra has appeared for five generations. Examine it. Doesn't it prove that pellagra is a hereditary disease? Does it not appear that there is a gene for pellagra which is passed on from generation to generation?

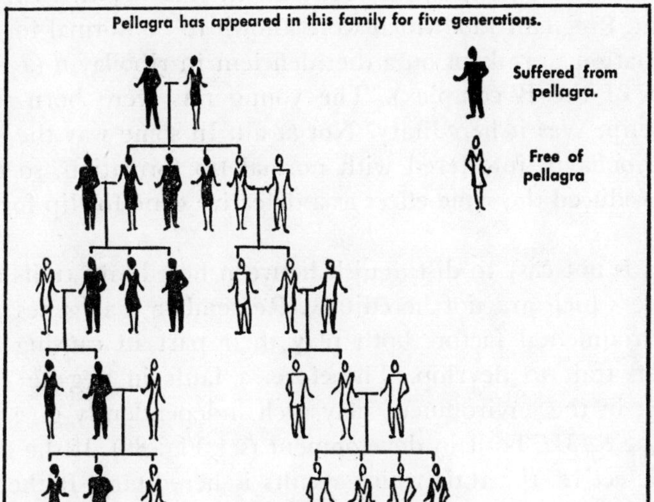

FIG. 85. Is pellagra inherited?

But wait. In 1915 Goldberger proved that pellagra resulted from the lack of a vitamin. He showed that pellagra was a dietary disease, not hereditary at all. Why, then, did it appear in five generations of the same family? Simply because for five generations this poor family had been eating the same deficient diet, and in each generation the deficiency showed itself in the form of the disease.

Let us consider another case. In the normal development of the embryo, the upper lip begins as three different sections which join together to form one lip. Sometimes one of the joints remains open so that the lip has a split in it. This condition is known as harelip. Suppose a child is born with a harelip. Is it correct to assume that because the condition is present at birth it is a hereditary condition? We know that in certain strains of rats, harelip is gene-controlled. If this condition is hereditary in rats, can it not also be hereditary in man?

Before you jump at any conclusion, consider this experiment. Pregnant rats which were known to be normal for lip formation were kept on a diet deficient in riboflavin (a vitamin of the B complex). The young rats were born with harelip. Was it hereditary? Not at all. In some way the lack of riboflavin interfered with normal lip formation, so that it produced the same effect as a defective gene for lip formation.

It is not easy to distinguish between hereditary traits and those which are not hereditary. Remember that genes and environmental factors both play their part in causing any given trait to develop. Therefore, a fault in a gene or a fault in the environment may each independently give rise to the *SAME* fault in development (see Fig. 86). If the gene is defective, the trait which results is hereditary. If the environment is defective, the trait which results is not hereditary, yet outwardly it may appear identical to the hereditary

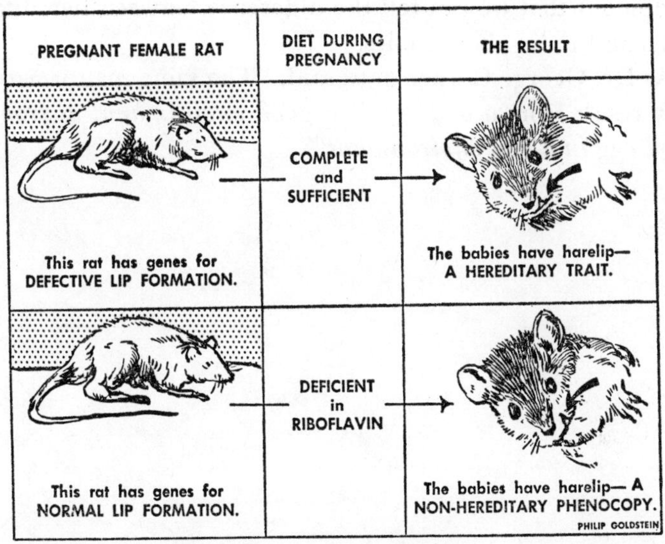

FIG. 86. Is harelip inherited?

trait. An environmentally caused trait which is a copy of a hereditary trait is called a PHENOCOPY.

So beware of jumping to conclusions. Do not assume, when a mentally defective child is born, that the family tree is tainted by a defective gene. It *may* be that the defect is hereditary, but it is just as likely that it is not. Perhaps the mother was infected with the virus of German measles during the first two months of pregnancy. Perhaps she was exposed to X rays in early pregnancy without proper precautions. Perhaps her diet was deficient in some nutrient necessary for proper development of the embryo's nervous system. Perhaps she suffered from oxygen deficiency at the critical period when the brain of the embryo began developing. These are only a few of the many environmental factors which might cause a child to be born with mental defects. We are not yet so far advanced in our understanding of development that we can tell the difference between hereditary traits and their phenocopies.

As Dr. George Corner once said, "The same malformation may result from *a bad egg in a good environment as from a good egg in a bad environment.*"

CHAPTER 52

WELL-BRED PLANTS AND ANIMALS—THE IMPROVEMENT OF PLANTS AND ANIMALS

EVER SINCE man began to domesticate plants and animals, he has been trying to improve them. From the start it was observed that the animals or plants of every kind differed among themselves. Since some of these variations seemed to be more satisfactory than others, the farmer or herder tried to eliminate those he considered inferior, while he wanted those with the desirable variation to increase. There is no doubt that over the centuries a gradual improvement in crops and herds did take place. But progress was slow—very slow—until the science of genetics was established, because with an understanding of heredity came an understanding of how improvement could be brought about.

How far we have progressed in the fifty years since genetics became a science is illustrated by the following quotation from *An Account of the Activities of the Agriculture Research Center at Beltsville, Maryland,* issued by the United States Department of Agriculture in 1948.

Modern plant breeders are not satisfied with selecting good types, getting them to breed true, and using them to replace old ones. Rather, they formulate an ideal in their minds, and proceed to create something that meets this ideal as nearly as possible by combining the genes from two or more plants. Their ambition does not stop at getting a plant with a single desirable characteristic. They strive for many desirable characteristics in one plant —for example high yield, high quality, and resistance to disease.

The two sections which follow will outline briefly what breeders aim to accomplish, and how they hope to achieve their objectives.

CHAPTER 53

THE AIM OF BREEDING

GENERALLY SPEAKING, the aim of breeding is to bring about improvement in animals and plants. But let us be clear on what we mean by improvement. We are looking at things from the point of view of man himself—not from the point of view of the organism being improved. What man considers a better animal is not necessarily a better animal in the wild state. Thus, a heavier pig may appeal to a farmer because of the greater amount of meat it carries, but the

Is *longhorn* a desirable trait?

heavier pig would certainly be at a disadvantage in the wild, in competition with wild pigs. When we speak of improvement, we mean improvement for the benefit of mankind. Keeping this in mind, we can now discuss some of the specific ways in which man likes to improve his domesticated organisms.

First, it is to man's advantage if production is high. Thus, he breeds for cows that give more quarts of milk; for corn or wheat which produces more grain per acre; for chickens that lay more eggs; for larger strawberries or blueberries; for sugar beets with higher sugar content; for pine trees which grow faster to replace those cut down. In general, he tries to make two blades of grass grow where one grew before —he tries to increase the amount produced.

Second, it is desirable to produce a product of the best quality possible. So he breeds for steers with tenderer meat; for cows which give creamier milk; for corn with more oil (or sometimes with less oil); for berries with sweeter taste; for flowers with brighter colors. In general, he tries to improve the quality of what he is producing.

Third, it is of benefit to man to develop animals and plants which do not suffer the ravages of disease. Under cultivation, plants and animals live close together in large numbers. This creates ideal conditions for the spread of epidemics. Thus, a field of wheat may be devastated by black stem rust; or thousands of tomato plants may suddenly wilt after the grower has put in countless hours of work; or a herd of cattle may fall victim to hoof-and-mouth disease; or a swarm of bees may be completely destroyed by American foulbrood; or the virus of curly top may sweep through a field of sugar beets, making it necessary for the crop to be abandoned. Therefore, the breeder tries to develop varieties which are immune to these diseases—wheat immune to rust; tomatoes resistant to wilt; cattle immune to Texas fever;

chestnut trees which will resist the blight; bees immune to foulbrood; sugar beets immune to curly top.

Fourth, there is a limitless field for the breeder pertaining to special traits which he considers desirable. For example, turkeys are normally big birds—too big for the average small family. This was remedied by the development of the Beltsville turkey—small size, but with lots of white meat. Race horses are bred for speed; truck horses for pulling power. Flowers are bred for showiness, for large bloom, for double flowers. Lawn grasses are bred to withstand hot, dry weather, and to crowd out crab grass—that ever-present threat to the beauty of the lawn. Watermelons are bred to refrigerator size. Dwarf varieties of trees are bred for the home gardener. Broomstraw is bred long for regular brooms and dwarf for use in whisk brooms. Bees are bred for long tongues so that they can reach for the nectar in deeper blossoms.

Fifth, the breeder is always on the lookout for new varieties. There is always a market for a new kind of apple or a new breed of sheep, and some of the new varieties may prove to be better than the old.

CHAPTER 54

THE METHODS OF THE BREEDER

A. Selection—Picking the Best

How does the breeder go about his work? His basic method is so old that we do not know where or when it began. It is a method known as SELECTION, which means simply picking out what you think is best. When you want a new suit or a new dress, you go to the store and select *the one you like* from a rack which holds dozens of dresses or suits. The breeder does likewise. He observes his field of blueberries, and says, "This plant I like. It bears big berries which taste good." Or he says, "This chicken lays more eggs than the others. There is a good market for chickens that lay many eggs."

And then he does an important thing. He cuts down every berry plant in the field except the one or two which he selected. Or he sells to the butcher all the chickens in his hen house except the few he has picked out. He does this to prevent the ones he considers undesirable from producing offspring. He breeds only the few he selected for this purpose, on the theory that "like begets like." If this chicken is a good egg producer, it is likely that her children will also be good egg producers, and if a chicken is a poor egg producer, it is likely that her children will also lay few eggs.

But this is where the method of selection falls down, for it is not always true that "like begets like." Differences as well as similarities can be inherited. Two black rabbits can produce a white rabbit in the normal course of passing on hereditary traits, and two white-flowering sweet peas may produce a purple one. Similarly, hens selected because they lay many eggs may pass on genes for poor egg production to some of their children.

Therefore, selection of a more effective type is practiced by the breeder who knows his genetics. He carries on what is called PROGENY SELECTION. If he wants cows that give more milk, he selects for breeding only animals whose *children continue* to produce milk in large quantities. In this way he knows he is breeding animals which can pass on only the desirable genes. Even by this method, selection cannot bring out anything which is not already in the genes. All selection can accomplish is a concentration of the "good" genes. Selection can produce a herd in which more animals carry the "good" genes, *but it can never make the genes "better."*

B. Inbreeding—Exposing the Hidden Traits

Once the breeder has selected the animals or plants which he considers to be desirable, he wants to establish a strain which is pure for this trait. If he knows genetics, he realizes that undesirable traits may be hidden as recessives in some of his stock, and that, in time, these undesirable traits will inevitably crop out. So his objective is to eliminate the plants or animals which carry such traits. But how can he tell? He cannot see the undesirable hidden genes, and therefore he must resort to a method called INBREEDING. This simply means the mating of closely related plants or animals.

In many plants the problem is simplified because it is possible to self-fertilize. Thus, when the improvement of corn was begun by skilled geneticists, their first task was to inbreed for many generations. Their corn got steadily worse, but they continued. They were exposing all the undesirable traits which were hidden in the plants. They continued until they developed PURE LINES—strains of corn which always remained the same even when inbred for many generations.

In animals, this is much more difficult, because self-fertilization is impossible. Here the breeder must resort to the crossing of close relatives to get the same effect. Thus he may BACKCROSS a hen to her father, or a young rooster to his mother. Or he may allow a male and female from the same litter of pups to mate when they grow up. Or he may cross a young bull with a cow which had the same father as the young bull.

You may ask at this point, "Isn't that dangerous? Doesn't such close breeding weaken the stock?" The answer is, "Maybe!" Remember that the purpose of inbreeding is to bring out the hidden traits. If there are hidden traits which are bad, they will most certainly be brought out, and the stock will be weakened. But suppose all the bad traits have been weeded out. Suppose that there are only desirable traits —even the recessives. Now inbreeding, no matter how close, cannot bring out anything bad. On the contrary, it serves to fix the desirable traits and even intensify them. One more fact is of interest, as evidence that inbreeding in itself is not harmful. Peas, beans, oats, barley, rice, and many other plants are naturally self-fertilizing. Although they have been practicing this closest possible type of inbreeding for thousands of years, there is no loss of vigor or fertility.

C. Crossbreeding—Combining Traits

Suppose that, by constant inbreeding, a breeder has developed several pure lines of corn. One of these has good ears, another has sturdy stalks, and a third is immune to smut. The breeder would like to develop a single strain of corn which combines all three traits. Can this be accomplished?

Again the answer is "Maybe!" If the breeder has all the time in the world, and all the money in the world, and if he is lucky, he can probably accomplish this aim by crossbreeding. CROSSBREEDING, or HYBRIDIZATION, is the mating of unrelated organisms. He starts by cross-fertilizing two of the pure lines. When the new generation is produced, he selects the most likely of these—the ones which show the two traits he wants—and he inbreeds them. If he can establish a strain pure for the two traits, he then crosses this with the line pure for the third trait. This is a long-drawn-out process which requires a great deal of advance planning and good judgment in selection. But it also needs the assistance of a little good fortune, since the recombination of traits is strictly a chance process (see chap. 12).

There is an old joke which illustrates this. It tells of a "mad scientist" who crossed roses with onions because he wanted to develop a strain of onions which smelled like roses. But alas! All he could get was a rose that smelled like onions. Bad luck made him a "mad scientist," whereas if he had been lucky he might have become a genius.

Here is an actual case. Tomato growers have long been plagued by a disease called fusarium wilt, which destroyed whole fields of tomato plants. A world-wide search for immune tomato plants led to the discovery, in Peru, of a wild relative of the tomato which was completely resistant to the disease. But its fruits were only about 1/4 inch in diameter, a far cry from the kind of tomato which the American housewife will buy. In 1936 the geneticists at Beltsville began an intensive program of hybridization and selection. Their aim was to combine into a single plant the resistance to fusarium wilt of the tomato from Peru, with the large size of the American Marglobe tomato. By 1940 they had achieved this objective, and they were able to release to growers the Pan American—a tomato of large size and high fusarium-wilt resistance. Good planning, skillful management, knowledge, and good luck all joined to give us a better plant.

Sometimes hybridization occurs between animals or plants of different species. Some of these wide crosses produce useful animals such as the mule. On the other hand, some, such as the cross between the radish and cabbage, are of interest to geneticists only. The cattalo is a hybrid produced by crossing the buffalo with cattle of the domestic breeds. A comparable cross has been made in Asia between the yak and domestic cattle. A tiger-lion hybrid called a tiglon was on exhibition in the New York zoo. And a zebronkey, resulting from a cross between a zebra and a donkey was shown in the zoo in Manila. There are stories of matings between coyotes and dogs in upper New York State, and some of the best sled dogs in Alaska are supposed to be wolf-dog hybrids.

Frequently, hybrids are bigger, stronger, and more vigorous than either parent. Hybrid corn is well known for its great yield, and the mule is stronger than the horse or jack. An interesting example of hybrid vigor is known to fishermen. The hybrid of a muskellunge and a northern pike is called the tiger muskie because it is a better fighter than either of its parents. However, hybrids of such wide crosses are often sterile, so that it is difficult to establish breeding stock.

D. The Problem of Linkage

As you recall (see p. 66), in linkage, traits are inherited as a group because their genes are carried on the same chromosome. This can be a boon to the breeder under certain circumstances, or it can be a thorn in his side. The imaginary cases cited below will explain.

Wouldn't it be wonderful if every cow carrying genes for high milk production also had a black spot between the eyes? If this were so, the breeder could pick out at birth those calves which would later produce large quantities of milk. All calves which did not have the black spot could be butchered and only the good milkers raised and fed. Unfortunately no such linkage actually exists. But in certain breeds of chickens, males can be distinguished from females, shortly after hatching, by a sex-linked feather pattern. This is of great advantage to the egg farmer, who does not like the idea of feeding young roosters because they do not pay off in eggs. Yet, if he cannot tell the hens from the roosters at hatching, he must feed them all for several weeks until sex can be recognized.

But linkage may cause trouble, too. Suppose the breeder would like to develop a variety of large blackberries on thornless stems, but he finds that thornlessness is linked to small berry size. Try as he will, he cannot separate the two traits, and so he is stuck—he must accept the thorns if he wants large berries. In this case, linkage prevents the breeder from achieving his purpose. In an actual case, Canadian barley breeders found that resistance to stem rust was linked with low kernel weight. Therefore they were unable to develop rust-resistant barley with large kernels until the linkage was broken by chance crossing over.

E. Mutations and the Breeder

One thing we have learned is that breeding and selection cannot produce anything new. They can develop new combinations of traits, or they can bring out hidden traits, but they can never produce new traits. This is because selection and breeding can only reshuffle the genes which are already present. No new genes can be created.

But we have also learned that mutations are always occurring (see p. 99). These mutations give the breeder new material with which to work, and so the breeder is constantly on the lookout for changes. Some of these mutations become the basis for new varieties. For example, the pink grapefruit originated as a mutation in a single branch of an ordinary grapefruit tree growing in a grove in Florida. Platinum fox and silverblu mink were mutations which breeders discovered and developed. The Golden Queen raspberry originated when a breeder realized the market value of this cream-colored mutant that grew on part of a plant which normally produced red raspberries. In each case, a new type appeals to some breeder, who then develops it into a new variety.

However, breeders do not always sit and wait. They may help nature along a little. By using X rays, neutrons, mustard gas, etc., they speed up the rate of mutation. Of course, they cannot produce the particular mutation they want, but where there are more mutations to choose from, there is more apt to be one which is desirable. Using this technique on the mold *Penicillium,* many mutants were produced. Among them were some with far greater penicillin-producing power than the original had. These were selected and cultured.

F. The Breeder and Environment

The breeder who understands genetics realizes that heredity merely gives an organism a capacity to develop certain traits *providing the environment is right.* He knows that it is pointless to plant hybrid corn in a waterless desert or in a rocky field lacking good soil, for the return will be poor. The corn will not provide the large yield which its genes indicate that it can. It is just as foolish to put prize cows on a starvation diet. They will never produce large quantities of rich, creamy milk, even though they have the best genes. A pig with genes for large size will still be a runt if he is infected with worms, and a rosebush which is attacked by aphids will never be able to show the beautiful blossoms it is capable of producing.

The breeder does not permit the capabilities of his plants and animals to be smothered by poor environment. He tries to provide the best of everything. He plants his corn in good soil, and adds fertilizer to boot. He wants to give those genes the best chance to show what they can do. He feeds his cows the best diet he knows of, but he also carries on experiments to find even better diets, because he knows that a better diet will result in more milk, even with the same genes. He sprays his roses to destroy the aphids in order to give the rosebush the chance it needs to bloom. And he keeps his pigs under conditions so sanitary that worm infection is unlikely.

It is true that the breeder tries to develop plants and animals with good heredity, but he also tries to give them the best possible environment in which this heredity can develop.

CHAPTER 55

HUMAN HEREDITY AND THE IMPROVEMENT OF THE HUMAN RACE

AND WHAT about man? Can man also be improved? If we can produce better plants and animals through proper application of our knowledge of genetics, can we also use the same knowledge to develop a superior race of humans? Here is a question which is truly worth thinking about. But don't jump at any conclusions!

"Certainly man can be improved!" you might be tempted to say. "Man is an animal like the others. Since he functions the same way, his heredity must follow the same genetic rules as that of any other animal. So, it ought to be easy to apply to the human race the methods of the breeder, and produce better people in the same way that we produce better turkeys or better corn."

On the other hand, your first reaction might be, "Oh sure, the human race could stand some improvement. But man is different. He's an animal all right, but a different kind of animal. Man is an animal with an intellect, with a soul. He's a reasoning animal, and you cannot treat him as you would a sheep or a cow. So you can't apply the same methods to man that you would use on another animal."

Which view is correct? There is room for argument on both sides, and there are many questions to be settled before any reasonable conclusion can be reached. For example, we would want to know whether there is any evidence that human heredity does, in fact, follow the same laws as does heredity in other animals. In order to answer this question, we would first have to know how scientists study human heredity, and what they have learned from their studies.

Next we would have to decide what we mean by a better human race. If we are trying to improve the human race, we must have a clear picture of what it is we are striving to attain. Only after we understand our objective can we begin to think of methods by which this objective can be achieved. Then, when we understand how human heredity operates, when we know clearly what our goal is, we can begin to consider whether the breeder's methods are applicable to man—whether these methods have ever been tried, and with what success.

But we must begin at the beginning. Therefore, the next section will consider the methods which are used to study human heredity, and what has been learned by using them.

CHAPTER 56

HUMAN HEREDITY—HOW IT IS STUDIED

IT WOULD BE nice to know whether human traits are inherited in the same way as are the traits of other organisms. But we still face the problem of finding out if this is so. You may ask, "Why is this so difficult? Why don't we apply Mendel's method (see p. 113) to the problem? This method has proved to be very effective in analyzing the heredity of all sorts of plants and animals. Why should it not be effective in man?" But Mendel's was an experimental method. Can an experimental method be applied to the study of heredity in human beings? Let us consider a practical situation.

A certain scientist wishes to study the inheritance of eye color in *Drosophila*. He starts with a strain of flies, all of whose ancestors have had red eyes, and another strain, all of whose ancestors were white-eyed. He sets up a hundred milk bottles with the proper food, and into each of fifty bottles he puts a red-eyed male and a white-eyed female. In the rest of the bottles, he reverses the situation by using a white-eyed male and a red-eyed female. Now he stores the bottles in convenient racks in his laboratory where all the external conditions—light, heat, moisture, etc.—are carefully controlled.

In a few weeks he is rewarded with a crop of two or three thousand first-generation flies. He observes their eye color, and sets up another hundred bottles each containing an F_1

male and female. A few more weeks pass, and he can count the thousands of F_2 offspring to see how many have red eyes and how many have white eyes. Thus he determines how the trait segregates. In a matter of a few months, the scientist has learned much about the heredity of eye color in *Drosophila*. He may even be in a position to reach a valid conclusion, because he has carefully set up an experiment in which all possible conditions have been controlled; he has repeated the same experiment a hundred times; he has several thousand cases upon which to base his conclusions.

Now consider the plight of another young scientist who is studying the same trait, eye color, in man. He cannot select a blue-eyed man and order him to marry a particular brown-eyed woman. No indeed. He cannot do this even once, much less a hundred times. So he is forced to depend on luck. Instead of carefully setting up the experiment, the scientist must hunt around for married couples in which one member happens to be blue-eyed and the other brown-eyed. Of course, he does not know if the ancestors of the blue-eyed persons have always had blue eyes. Nor does he know about the ancestry of the brown-eyed people. Immediately, he realizes that this can never be a perfect experiment, but since he knows he cannot hope for perfection, he is ready to go on. He hunts around until he finds one hundred couples which fit the needs of his study.

However his troubles have only begun. He must now wait for these couples to have children, and that takes time. At the end of ten years, his hundred couples have produced some two hundred children. But were all these children born and raised under the same conditions? Of course not. Some were rich, some poor; some were well taken care of, others neglected; some had good food, others had an inadequate diet; some had measles or diphtheria, while others had chickenpox or tuberculosis; some lived in the city, some in the country. The scientist had absolutely no control over the conditions existing in the homes where these children grew up. Did these diverse conditions have any effect on the inheritance of the trait being studied? He can never know.

And now comes another problem. How does the scientist get the F_2 generation for his study? The children will not be ready for marriage for another ten or fifteen years—and then of course it is very unlikely that they will marry among themselves. But even if they do, it takes another ten years for the F_2 generation to come along. This makes a total elapsed time of twenty-five or thirty years since this very unsatisfactory experiment began. And even after all this time, the scientist, now aged fifty or sixty, cannot really arrive at a satisfactory conclusion because he could neither control the conditions nor gather enough cases to make a valid generalization.

Obviously, the study of human heredity by the experimental method is out of the question, and another approach must be devised. The next best thing is a statistical attack on the problem. This means the collection of as much information as possible on the inheritance of certain traits. Such information is collected by means of pedigrees (family histories), through questionnaires, by direct observation, by reference to medical records, etc. The information so obtained is classified, tabulated, and analyzed, and, from the analysis, some conclusions can be reached. Unfortunately, much of the data collected in this way is not accurate, for a variety of reasons.

First of all, people are not anxious to reveal family secrets. This is particularly true for traits such as mental defects, which are considered undesirable. Most people prefer to keep such family skeletons hidden from view in the closet.

Second, people don't remember accurately. Just test yourself by trying to remember the eye color or hair color of your grandparents. It is unlikely that you could give accurate information about their traits.

Third, people do not interpret things in the same way.

The word "tall" does not mean the same thing to all people. What is dark-skinned to one person may be light to another. There are few standards against which to compare the trait being observed.

Fourth, the pedigrees which are pooled were collected by different investigators, each with his own methods and his own point of view. There is no guarantee that the data in all these pedigrees are actually comparable, even when all the pedigrees are supposed to deal with the same trait.

Fifth, there is no guarantee that the pedigrees which have been collected are really a sample of the population as a whole. For example, suppose we are studying the inheritance of hemophilia. We can include in our data all families in which hemophilia actually appears, but, since hemophilia is a recessive trait, the gene is carried in hidden form in many other families. These families are excluded from our data, even though they are transmitting the gene which we are trying to study.

Sixth, there is no guarantee that every case of a given trait is caused in the same way. In corn there are over a dozen genes each of which can produce albinism. Yet if we were studying albinism in corn by the statistical method, we would lump all these cases into one group, and we would get a false picture of the inheritance of albinism. Similarly, if we pool all pedigrees which deal with harelip (for example), we may be pooling some which are caused by heredity and others which are environmental in origin.

Seventh, different investigators can interpret the same data to mean different things. It is a well-known fact that statistics can be made to say almost anything one wants them to say. Thus, the social or political views of the investigator, or his early training, might influence his interpretation of the data, without his even realizing it.

Despite all these drawbacks, the statistical method has enabled us to learn a good deal about human heredity. Most of our information is drawn from family pedigrees, from medical records, from prison records, from the study of twins, from the study of random populations, etc. And all of these studies seem to point to the conclusion that heredity in man DOES follow the same basic pattern as does heredity in any other living thing.

We find evidence in human heredity of dominance, blending, segregation, recombination, independent assortment of traits, linkage of traits, multiple allels, multiple-factor inheritance, sex linkage, mutations, etc. Everything we find in the inheritance of plants and animals is also found in human inheritance.

So, we are ready to go a step further in our search for an answer to the question, "Can our knowledge of heredity be used to improve the human race?" We must now decide upon our objective, and that leads us to raise the question, "What is a desirable human being?"

CHAPTER 57

GOOD AND BAD—WHAT IS A DESIRABLE HUMAN BEING?

SUPPOSE YOU try to analyze one of your friends. You go about it in a systematic way, and you set down his traits—both good and bad—as follows:

Physical traits. He is tall and well built. He has broad shoulders and well-developed muscles. His eyes are dark, and what is left of his hair is black. His sight and hearing are normal, and he is rarely ill. On the negative side however, his features are spoiled by a large hooked nose. He became bald at the early age of twenty-two. He is red-green color blind, and he suffers from ulcers every time he has a business reverse.

Mental traits. He is endowed with high intelligence and is especially good at figures. He is very clever, and he reasons well. He is a keen observer. On the other hand he has no artistic, literary, or musical ability, and he tends to be rather nervous and jumpy at times.

Character traits. He is a responsible man in his community, reliable and trustworthy. He is a reverent man, who goes to church regularly and gives large sums to charity. He is temperate and thrifty. He lives for his wife and family. By contrast, he does not hesitate to be dishonest in his business dealings. Nor does he hesitate to take advantage of the weaknesses of others. He is very opinionated, and his mind is closed to other viewpoints. He is very susceptible to flattery.

Personality traits. He is a friendly man, who can be kind and helpful when he wants to. He is usually cheerful and pleasant—a good social mixer. He has good manners, and he is usually considered to be a good sport. He is punctual in keeping appointments. By contrast, he is very blunt and undiplomatic. He is prone to make nasty remarks about people. He is inconsiderate and intolerant of the rights of anyone who crosses him. He is reactionary in his political views.

What conclusion do you reach? Is he good or is he bad? Is he, or is he not a desirable human being? Obviously, he is a mixture of both. He has many desirable traits, and just as many undesirable ones. So the first conclusion we must reach is that there is no *good* human and there is no *bad* human. Every person is a little of each.

Here is an important thing to keep in mind. Every person has some desirable traits. While the same is true of plants and animals other than man, the breeder is willing to sacrifice everything for the *single trait* in which he is interested. He is willing to throw away a good trait if it is tied to a bad one. When we are dealing with humans, however, we are not ready to do the same. Would you, for example, throw away the electrical genius of a Steinmetz because he was a hunchback; or the literary genius of a Byron because he had a crippled leg? Of course not. The world could ill afford this kind of waste of human resources.

Now let us analyze a step further. How many of the traits which you listed for your friend are desirable? He is tall. Is tall desirable? How tall? Would you say that 6 feet and 11 inches is desirable—or is that too tall? Perhaps if we asked an African pigmy we would get a different answer as to what height is desirable in a person.

Do you like your friend's dark eyes? Do you think this is a desirable trait? Or, do you feel that this trait is neither desirable nor undesirable—it just doesn't matter. Yet not too long ago, Hitler rose to power on a program which claimed that blue eyes and blond hair were the marks of a superior race.

So you see that what is desirable to you may be unimportant or undesirable to another human being. Who is to decide which is correct? Who has the right to set himself up as judge? Naturally, we can all agree about certain extreme traits such as insanity or blindness. Certainly, they are undesirable. By the same token, we may all agree that great musical ability or a normal body are desirable. But for the large bulk of human traits, there is no universal agreement.

But that isn't all. Not only do different groups of people have different concepts of what is desirable, but the same group may change its mind from time to time. Styles in men and women change. While one generation of Americans may like its girls to be slender and thin-waisted, another generation may like them rather plump and buxom. What we consider to be a desirable trait today may have been thought undesirable a few hundred years ago. And who can foresee what our ideal man will be like a few hundred years hence.

Perhaps we can summarize in this way. Every person is a mixture of good and bad traits. But what is a good trait—and what is bad? There is no universal agreement on this. Our views differ from time to time and from place to place. Still, should we give up? Should we throw up our hands in despair? Should we say that improving the human race is a completely hopeless task? Certainly not! We can set up certain goals, even though we are not sure that our descendants will want to continue to aim at these same goals. We can strive to achieve the goals we set for ourselves by whatever methods are at our disposal. Therefore, the next section will consider two approaches to the problem of race improvement.

CHAPTER 58

RACE IMPROVEMENT—TWO APPROACHES

THE PROBLEM of race improvement is not new. History tells us of various attempts to accomplish this. For example, it is said that the Spartans threw over a cliff every infant which showed a defect at birth. And certain tribes of Indians are supposed to have dipped the newborn child into a stream of icy water on the theory that if he could not survive this treatment he would never grow up to be a desirable man. On the other hand, Napoleon and Hitler each tried to develop special groups of supermen by selective marriages.

Our usual approach to this problem is less drastic. Actually, there are two different approaches—the EUGENIC and the EUTHENIC. Both eugenics and euthenics hope to achieve the same goal—namely, the improvement of the human race—but each employs a different method.

One might imagine "Mr. Eugenics" saying, "There are many undesirable people on the earth today—insane, feeble-minded, crippled, criminal, disfigured, etc. These undesirable people continue to produce undesirable children—remember, like begets like. As a matter of fact, these undesirables have more children than do the desirable people—the intelligent, hard-working, capable, perfectly formed people. And what is the result? The pool of human germplasm is gradually deteriorating. More and more poor germplasm is poured into the pool, while proportionately less and less good germplasm is put into it.

"We must put a stop to this. We must reverse the trend. We want the pool of human germplasm to get constantly better in each generation—not worse. We want to increase the amount of *GOOD* germplasm and decrease the amount of bad germplasm in the pool. This isn't so difficult. All you have to do is prevent the undesirable from having children, and thus prevent them from passing on their undesirable traits; on the positive side, you can make it possible for the desirable people to have more children, and thus increase the amount of good germplasm being passed on."

On the other side, "Mr. Euthenics" might say, "True, there are many undesirable people on the earth. But we can't do very much about their heredity. Heredity is something which we cannot touch. But the environment—that's another proposition. We can do something about that.

"When a child is born into this world, it can be molded and changed by the environment. The surroundings in which the child grows up, its training, its education, the experiences to which it is exposed—these can make the child into an undesirable or a desirable person, as the case may be. So, if we want a better race of humans, we must give humans the best possible environment in which to develop. We must take the wonderful germplasm with which the human race comes equipped, and give this germplasm the best and most favorable conditions under which to develop. In this way, we will get better people and a better human race."

These are the two approaches. The eugenic approach would weed out the bad germplasm by some form of selection—some method of breeding. The eugenic approach would increase the amount of good germplasm in the human pool, and decrease the amount of bad germplasm. On the other hand, the euthenic approach would take human germplasm as it comes and give it the best possible environment in which to develop. In this way, the best can be brought out—the best in the good germplasm, and the best in the inferior germplasm. The eugenic approach would breed better people, while the euthenic approach would try to make all people better, regardless of their hereditary endowment.

Which method is better? Which is more practical? Let us first consider it theoretically. Let us assume that we can decide which traits are desirable, and that we have absolute control over marriages. Let us assume a society which is so constituted that it can say who is to have children and who is not to have any; that it can sterilize the unfit and subsidize the fit. Under these conditions, could a eugenic program actually improve the human race? Let us assume the extreme situation in which we can treat reproduction in people as the breeder treats it in cattle or sheep. Under these circumstances, it would seem that what has been achieved with animals should be possible with man. If we can produce cows which are superior milkers, we should be able to breed men who are superior thinkers. If we can breed flowers of breath-taking beauty, we should be able to breed a human race in which each individual is an Adonis.

Theoretically, we should be able to do it. But practically it won't work—at least not in our present society. We believe too much in the inalienable rights of the individual. We are not a society of robots or automatons, but a society of people, each with his virtues as well as his faults—each with the right to have a mate of his own choosing. True, we often freely give up certain of our rights for the good of society, and sometimes the rights we give up do fit in with the aims of a eugenic program. Several states in the United States, for example, have laws which permit the sterilization of the criminally insane or the hopelessly feeble-minded. Other states segregate them in order to prevent them from having children.

No—a full-fledged eugenics program is still too extreme for our society. Even if we were ready for such a program, we have no guarantee that it would be effective. Suppose we *could* sterilize every person who showed a defective trait, would the next generation be better? This is highly improbable. Remember that most of us, who are perfectly normal in every obvious way, actually carry hidden recessive genes for a variety of defects. These defects will continue to crop out, generation after generation. In addition, there is the matter of mutations. Mutations occur spontaneously, and frequently they cause defective traits to appear in families where they did not exist before. Unfortunately, we do not know any way of preventing genes from mutating.

And there is one more thing. How do you measure the value of a human trait? Consider this example, if you think you know how to do this.

Sickle-cell anemia (see section 27) is certainly a terrible affliction. People who inherit the sickle-cell gene from both parents suffer a painful death at a relatively young age. Certainly the sickle-cell gene is "bad" and should be eliminated from the pool of human germplasm.

But wait a minute. Before you decide to eliminate this gene, consider the experiment that Dr. Anthony C. Allison carried out in 1954 with human volunteers.* This experiment demonstrated that *persons carrying one gene for sickle-cell were far more resistant to malaria than persons with two genes for normal hemoglobin.* For a person living in a place where malaria is prevalent, it may even be an *advantage* to have a sickle-cell gene. It may very well save his life. So perhaps we should not be so ready to eliminate the sickle-cell gene from the pool of human germplasm unless we can first eliminate the threat of malaria.

---

* A more complete description of Dr. Allison's interesting experiment will be found on pages 52-59 of the book *TRIUMPHS OF BIOLOGY* by Philip Goldstein, Doubleday and Co., New York, 1965.

Perhaps, in centuries to come, when society has evolved to a higher level, and when our knowledge of human heredity is much further advanced than it is today, a eugenics program will be able to produce a race of superhuman beings. But not yet. With our present society and our present genetic knowledge, a full-scale eugenic program is impractical.

However, the euthenic program has immediate possibilities and can bring immediate results, for it does not plan to tamper with the genetic make-up of the individual or the group. In fact, the euthenic program points out that it is *unnecessary* to tamper with genes to get better people.

Consider the following very incomplete list of desirable human traits: braveness, charity, cheerfulness, consideration, courtesy, culture, diplomacy, good manners, good physique, good sportsmanship, handsomeness, health, honesty, intelligence, open-mindedness, pleasantness, punctuality, responsibility, reverence, strength, temperance, tolerance, trustworthiness, etc. How many of these traits are hereditary? Are there genes for honesty, or courage, or diplomacy? Can you inherit tolerance, or open-mindedness, or reverence? Did the fertilized egg from which you developed have determiners for good manners, or charity, or punctuality? As you see, many of these traits are environmental in origin. Much of what makes a human being "good" comes from his upbringing, his training, his schooling, his environment. Then why bother the genes? Since these traits do not have a hereditary basis anyway, how can a eugenic program affect them? Obviously, it cannot, and it is only by controlling the environment that we can develop these traits.

What about those traits which do have a hereditary basis? The euthenics program stresses the theory that few human beings ever tap the full potential of their genes, because the environment in which they developed never permitted the maximum level to be reached. Take the case of intelligence.

It is pretty well established that intelligence is gene-controlled. Then how would you account for the results in the following experiment?

A group of school children in an underprivileged area was tested for intelligence. Then the group was put on a superior diet for several months, and at the end of this period the intelligence was measured again. An average rise of several points was observed. What does this mean? Is it reasonable to assume that the poor diet of these children was preventing them from approaching their genetic limit of intelligence, and that a better diet enabled them to come closer to this limit? Of course, it does not follow that the right diet will make every person into a mental giant, but it does seem to indicate that if every child had a satisfactory diet, some of our slow learners might become normal in intelligence, and better people because of it.

The basic idea of the euthenic program, is this: Put each person into that environment which will best permit him to develop to the utmost potential of his genes. It is easy enough to see how this works out for a trait such as physical build. It is fairly obvious that a baby with "six-foot genes" will develop into a six-footer if he is exposed to the proper environmental conditions. On the other hand, his twin brother, with the same "six-foot genes," will be prevented from becoming a six-footer if he is subjected to poor living conditions, poor food, poor health, etc. But, by the same token, a baby with "five-foot genes" cannot become a six-footer no matter how superior the environment.

This is the limitation of euthenics. It cannot create anything new. It cannot bring out that which is not already laid down in the genes. The heredity is the limiting factor, but the environment determines to what degree the limit is approached. To achieve the ideal, the two must work hand in hand.

In summary, we are not advanced enough in our knowledge of human heredity or in the level of our society to enter upon a full-fledged eugenic program to improve the human race. Nor have we yet tapped the full potential of mankind as it is represented by its present pool of genes. Much of this potential is now prevented from developing by unsatisfactory environmental factors. Furthermore, many of the desirable human traits, such as character and personality, are completely environmental and have no relation whatsoever to heredity. Therefore, at the present time a full-scale eugenic program is impractical, although we can adopt small fragments of such a program. Our present hope for making a better human race lies in providing the best possible environment for every individual, in order to bring out the best which is contained in the pool of genes from which each generation of humans develops.

FOR ADDITIONAL INFORMATION —A BIBLIOGRAPHY

Genetics Is Easy cannot possibly be a complete survey of the field of heredity. Many areas of the subject have been touched upon only superficially; and some have been omitted entirely. For more complete coverage, the reader is referred to other books like those listed below. The author wishes to thank Mrs. Gabrielle Edwards and Mr. William Weiss for their suggestions concerning some of the books to be included in the list.

For convenience, the bibliography has been arranged under the following headings:

1. Material of historic importance in the study of heredity.
2. Biographic notes about famous geneticists.
3. Nobel Prize lectures in the field of genetics.
4. Important research papers in the field of genetics.
5. Popularized explanations of heredity.
6. More comprehensive treatments of the subject.
7. Material on modern aspects of genetics.
8. Material on human heredity in health and disease.
9. Materials useful to students carrying on projects in genetics.

1. MATERIAL OF HISTORIC IMPORTANCE IN THE STUDY OF HEREDITY.

DARWIN, CHARLES. *The Origin of Species by Means of Natural Selection or the Preservation of Favored Races in the Struggle for Life*, 1859.

——— . *Variation in Plants and Animals under Domestication*, 1868.

These are both classics in the field. Comments are unnecessary. Of course, they are pre-Mendelian.

MENDEL, GREGOR. *Experiments in Plant Hybridization,* 1866. English translation—Harvard University Press, Cambridge, Mass., 1948.

This is Mendel's original paper announcing the fundamental principles of heredity. It is written in clear, concise style. High school biology students can profit from reading the original paper, rather than reading about it. Certainly every biology teacher should be familiar with it. Sections of this paper are reproduced in the following references listed elsewhere in this bibliography: Bateson—*Mendel's Principles of Heredity;* Goldstein—*Triumphs of Biology,* pages 127-147.

DE VRIES, HUGO. "Concerning the Law of Segregation of Hybrids." *Reports of the Paris Academy of Sciences,* 1900, Vol. 130, pp. 845-847.

CORRENS, CARL. "G. Mendel's Law Concerning the Behavior of Progeny of Varietal Hybrids." *Reports of the German Botanical Society,* 1900, Vol. 18, pp. 158-168.

TSCHERMAK, ERIC VON. "Concerning Artificial Crossing in Pisum Sativum." *Reports of the German Botanical Society,* 1900, Vol. 18, pp. 232-239.

These are the papers in which three different men, from three different countries independently announced the rediscovery of Mendel's work. The following references contain excerpts from these papers, translated into English: Brooklyn Botanic Garden Golden Jubilee Supplement of "Gentics"; and *Triumphs of Biology* by Philip Goldstein, pages 148-165.

BROOKLYN BOTANIC GARDEN. *The Birth of Genetics.* Golden Jubilee Supplement to Genetics, Sept. 1950, Vol. 35, # 5, part 2.

This supplement contains English translations of ten letters written by Mendel to Carl Nageli, the famous Swiss biologist, between 1866 and 1873. It also contains the original research papers of DeVries, Correns, and Tschermak announcing the rediscovery of Mendel's work.

BATESON, WILLIAM. *Mendel's Principles of Heredity.* New York: Cambridge U. Press, 1909.

The early work in genetics is described by one of the pioneers in the field. Contains a biographical sketch of Mendel in the appendix (pp. 327-334) and a translation of Mendel's paper (pp. 335-379).

MORGAN, THOMAS HUNT. *The Scientific Basis of Evolution.* New York: W. W. Norton & Co., 1932.

An account of the early work in the field by the first Nobel Prize winner in genetics.

GARROD, ARCHIBALD E. *Inborn Errors of Metabolism.* London: H. Frowde and Hodder, 1928. (Originally appearing as a series of four articles in *The Lancet,* during July 1908).

This little book came before the world was ready for it. The work of Garrod, described in this book, laid the groundwork for the entire field of biochemical and medical genetics.

DUNN, L. C. (Editor). *Genetics in the Twentieth Century.* New York, Macmillan, 1951.

Contains 26 essays by authorities in the field on the progress of genetics during its first fifty years.

———. *Short History of Genetics.* New York, McGraw Hill, 1965.

Although this book was published in 1965, it only traces the history of genetics from Mendel to about 1939.

2. BIOGRAPHIC NOTES ABOUT FAMOUS GENETICISTS.

ILTIS, HUGO. *Life of Mendel.* New York: W. W. Norton Co., 1932.

Just about the best biographical work on Gregor Mendel. Translated from the German.

STEVENSON, LLOYD G. *Nobel Prize Winners in Medicine and Physiology, 1901-1950.* New York: Henry Schuman, 1953.

Only two men were awarded Nobel Prizes in the field of genetics before 1950: Thomas Hunt Morgan in 1933 (pages 165-170) and Hermann J. Muller in 1946 (pages 238-243).

REIDMAN, SARAH R. AND ELTON T. GUSTAFSON. *Portraits of Nobel Laureates in Medicine and Physiology.* New York: Abelard-Schuman, 1963.
Chapter 11 deals with the Nobel Prize winners in genetics.

FORTUNE MAGAZINE, THE EDITORS OF. *Great American Scientists.* Englewood Cliffs, N.J.: Prentice Hall, 1961. (Originally appearing as articles in Fortune Magazine.)
The unfolding of American science is traced through the lives and work of American scientists. Part 4, The Biologists, deals with the work in genetics.

3. NOBEL PRIZE LECTURES IN THE FIELD OF HEREDITY.

All Nobel Prize lectures are printed in a publication called *Les Prix Nobel,* usually during the year following the delivery of the lecture in Stockholm. The lectures are often reprinted in other publications.

MORGAN, THOMAS HUNT—Nobel Prize in 1933—"The Relation of Genetics to Physiology and Medicine." Reprinted in *Scientific Monthly* for July 1935, and in the *Annual Report of the Board of Regents of the Smithsonian Institute,* 1935, pages 345-359.

MULLER, HERMANN J.—Nobel Prize in 1946—"The Production of Mutations." A brief abstract of this lecture may be found in Stevenson, L. G., *Nobel Prize Winners in Medicine and Physiology,* pages 238-243.

BEADLE, GEORGE W.—Nobel Prize in 1958—"Genes and Chemical Reactions in Neurospora." Reprinted in *Science,* June 26, 1959, Vol. 129, pp. 1715-1719.

TATUM, EDWARD L.—Nobel Prize in 1958—"A Case History in Biological Research." Reprinted in *Science,* June 26, 1959, Vol. 129, pp. 1711-1715.

LEDERBERG, JOSHUA—Nobel Prize in 1958—"A View of Genetics." Reprinted in *Science,* Jan. 29, 1960, Vol. 131 pp. 269-276.

KORNBERG, ARTHUR—Nobel Prize in 1959—"Biological Synthesis of Deoxyribonucleic Acid." Reprinted in *Science,* May 20, 1960, Vol. 131, pp. 1503-1508.

OCHOA, SEVERO—Nobel Prize in 1959—"The Enzymatic Synthesis of Ribonucleic Acid."

WILKINS, MAURICE H. F.—Nobel Prize in 1962—"Molecular Configuration of Nucleic Acids." Reprinted in *Science,* May 31, 1963, Vol. 140, pp. 941-950.

WATSON, JAMES D.—Nobel Prize in 1962—"The Involvement of RNA in the Synthesis of Proteins."

CRICK, FRANCIS H. C.—Nobel Prize in 1962—for elaborating the structure of the DNA molecule with J. D. Watson.

LWOFF, ANDRE; JACQUES MONOD; AND FRANCOIS JACOB—Nobel Prize in 1965—for studies dealing with the regulation of genetic mechanisms.

4. IMPORTANT RESEARCH PAPERS IN THE FIELD OF GENETICS.

BOYER, SAMUEL H. *Papers on Human Genetics.* Englewood Cliffs, N.J.: Prentice Hall, 1963.

A collection of experimental papers representing some of the most significant contributions in the field of human genetics. College level or advanced high school level.

GABRIEL, MORDECAI L., AND SEYMOUR FOGEL (editors). *Great Experiments in Biology,* Englewood Cliffs, N.J.: Prentice Hall, 1955.

Pages 225-280 include important papers on genetics. College or advanced high school level.

MULLER, HERMANN J. *Studies in Genetics.* Bloomington, Indiana, Indiana University Press, 1962.

This book includes a selection of important papers by the Nobel Prize winner, Hermann Muller.

PETERS, JAMES A. *Classic Papers in Genetics.* Englewood Cliffs, N.J.: Prentice Hall, 1964.

An excellent selection of classical papers in genetics. Of great value to serious students of genetics.

5. POPULARIZED EXPLANATIONS OF HEREDITY.

AUERBACH, CHARLOTTE. *Genetics in the Atomic Age.* London: Oliver A. Boyd, 1956; or Essential Books Inc., Fair Lawn, N.J., 1956.

A charming little book written for the layman by an expert. In lucid language, it sets forth the facts relating to the effect of radiation on man.

———. *The Science of Genetics.* New York: Harper and Bros., 1961.

A popularized rendition of basic genetics geared for the interested and educated layman. It includes both classical and modern concepts.

BEADLE, GEORGE W. AND MURIEL BEADLE. *The Language of Life.* Garden City, N.Y.: Doubleday & Co., 1966.

This book was written by Muriel Beadle who is not a geneticist. However, her husband, George Beadle, is a Nobel Prize winner in genetics. Together they make a fine team to produce a book for the layman which is both accurate and up to date.

FRANKEL, EDWARD. *DNA—Ladder of Life.* New York: McGraw Hill Book Co., 1964.

Written for high school students by a high school teacher of long experience. Understandable and clear.

GOLDSTEIN, PHILIP. *Triumphs of Biology.* Garden City, N.Y.: Doubleday & Co., 1965.

The original research papers of famous scientists are used to trace the history of important biological discoveries. Several of the discoveries are in the field of genetics—inheritance of sickle cell anemia; inheritance of color blindness; Mendel's work and its rediscovery; molecular, bacterial, and viral genetics.

HAFFNER, RUDOLPH E. *Genetics, The Thread of Life.* Columbus, Ohio: American Education Publications, 1962.

Each chapter of this paperback unfolds some aspect of genetics in a concise, simplified way. Chapter 3 on chromosomes, and chapter 6 on biochemical genetics are especially worthy of note.

MONTAGU, ASHLEY. *Human Heredity.* Cleveland, Ohio: World Publishing Co., 1959. Paperback edition—Mentor Books, New York, N.Y., 1959.

This is an excellent book, recommended for high school and college students as well as for the interested layman. It presents the principles underlying human heredity in readable and understandable terms, with strong stress on the sociological implications.

MULLER, HERMANN J. *Out of the Night.* New York: Vanguard Press, 1935.

This book by a Nobel Prize winner in genetics, expresses a view of what genetics could accomplish for the improvement of the human race if it were not for certain practical considerations. While it is an old book, published in 1935, the ideas expressed are as interesting and as important as they ever were.

RANDAL, JUDITH. *All about Heredity.* New York: Random House, 1963.

While it is obviously impossible for such a book to tell *"all"* about heredity, the information it contains is written simply enough so that even a junior high school student can understand it. Well illustrated.

SCHEINFELD, AMRAM. *The New You and Heredity.* Philadelphia, Penn.: J. B. Lippincott Co., 1950. (A revision of *You and Heredity,* first published in 1939.)

While this book is now about 15 years out of date, it is still one of the best books on heredity ever written for the layman. It is an excellent book for high school students, but it should be supplemented with more up to date material.

———. *Your Heredity and Environment.* Philadelphia, Penn.; J. B. Lippincott Co., 1966.

A new book by the author of *You and Heredity,* which is intended to replace the earlier books.

WALLACE, BRUCE. *Population Genetics.* (A BSCS Pamphlet). Boston: D. C. Heath & Co., 1964.

A simple and effective treatment of a complex subject.

6. MORE COMPREHENSIVE TREATMENTS OF THE SUBJECT.

GARDNER, ELDON J. *Principles of Genetics.* New York: John Wiley & Sons, 1964.

A well written and well illustrated reference text on the college level. It includes both classical and modern concepts of genetics.

LEVINE, R. P. *Genetics.* New York: Holt, Rinehart & Winston, 1962.

A brief but thorough survey of basic genetics on the college level. The emphasis is on modern genetic theory.

MADDOX, JOHN. *Revolution in Biology.* New York: Macmillan Co., 1964.

This stresses the story of proteins and their fate in cells. It gives a detailed account of the work on DNA, RNA, and protein synthesis.

MERRELL, DAVID J. *Evolution and Genetics.* New York: Holt, Rinehart & Winston, 1962.

An interesting book, with good illustrations which applies genetic mechanisms to modern evolution theory.

SRB, A. M., R. D. OWEN, AND R. S. EDGAR. *General Genetics.* San Francisco: W. H. Freeman & Co., 1965.

A very good basic text including a fine bibliography. There is a great deal about microbial genetics and biochemical genetics.

7. MATERIAL ON MODERN ASPECTS OF GENETICS.

AHERNS, E. H. (Editor). *Concepts in Biochemistry.* New York: The Rockefeller Institute, 1964.

This paperback is a collection of 10 essays written by scientists for the benefit of science writers who attended a briefing conference. In other words, scientists have interpreted various concepts in biochemistry for the writers, in the hope that the writers could then interpret better for the public. The first essay contains an excellent discussion of modern knowledge of DNA and of the way in which viruses are used for nucleic acid research.

Anfinsen, Christian B. *The Molecular Basis of Evolution.* New York: John Wiley & Sons, 1961.

A highly recommended treatise on biochemical genetics. However, it is intended for the advanced reader. Discussions are not only descriptive, but they also raise many problems that require additional research. Highly recommended.

Braun, Werner. *Bacterial Genetics,* 2nd edition. Philadelphia: W. B. Saunders, 1965.

A revision of a standard text on bacterial genetics.

Barry, J. M. *Molecular Biology: Genes and the Chemical Control of Living Cells.* Englewood Cliffs, Prentice Hall, 1964.

A paperback which gives an excellent analysis of the structure of genes and how they work. Discussion is at the molecular level. Rather difficult, but a good supplement to standard college texts.

No book on heredity can remain up to date for long. The field of genetics is changing too rapidly, and new discoveries are being made every day. Therefore magazine articles dealing with current research are an invaluable supplement to the book. *The Scientific American* frequently carries articles by authorities. These articles are usually well written and well illustrated. Reprints of many of the articles are available from the W. H. Freeman Publishing Co., San Francisco, California.

8. MATERIAL ON HUMAN HEREDITY IN HEALTH AND DISEASE.

Carter, C. O. *Human Heredity.* Baltimore, Md.: Penguin Book Co., 1962.

The author deals with diseases and malformations caused by various genetic factors. For college level.

Dobzhansky, Theodosius. *Evolution, Genetics, and Man.* New York: John Wiley & Sons, 1963.

This classic book first appeared in 1955. It is an excellent, authoritative discussion of the evolution of man based on modern genetic principles. Extremely well written and well illustrated.

GATES, REGINALD R. *Human Genetics.* (2 volumes) New York: Macmillan, 1946.

A basic reference book on human genetics covering what was known to the time of its publication.

NEEL, J. V. AND W. J. SHULL. *Human Heredity.* Chicago: U. of Chicago Press, 1954.

A fine book written at college level.

STERN, CURT. *Principles of Human Genetics.* San Francisco: W. H. Freeman and Co., 1959.

A standard work on human genetics.

SUTTON, H. ELDON. *Genes, Enzymes, and Inherited Diseases.* New York: Holt, Rinehart & Winston, 1962.

A small book on the college level. It uses the molecular approach to the inheritance of metabolic disorders.

WALLACE, BRUCE AND THEODOSIUS DOBZHANSKY. *Radiation, Genes, and Man.* New York: Holt, Rinehart and Winston, 1963.

A paperback that reviews the effect of radiation on living populations of organisms, including man.

WHITTINGHILL, MAURICE. *Human Genetics and Its Foundations.* New York: Reinhold, 1965.

This text on human genetics is not overburdened with statistical studies. It is an attempt to explain human inheritance through the careful exposition of fundamental principles. College level.

9. MATERIALS USEFUL TO STUDENTS WHO INTEND TO CARRY ON A PROJECT RELATING TO HEREDITY.

GOLDSTEIN, PHILIP. *How to Do an Experiment.* New York: Harcourt, Brace & World, 1957.

This is a general reference on project work, written by a high school teacher for high school pupils. It outlines the basic steps involved in carrying on an experimental project, from the selection of a problem to the final written report of results and the oral presentation before an audience. Highly recommended.

BIOLOGICAL SCIENCE CURRICULUM STUDY. *Research Problems in Biology* (4 series). Garden City, N.Y.: Doubleday & Co., 1963 and 1965.

Each series contains 40 research ideas for superior high school students. They are open-ended, so that the research may go off in many different directions. Some of the ideas relate to research problems in heredity:

Series 1 (1963) projects # 2, 15, 21, 25, 29.
Series 2 (1963) projects # 45, 58, 70, 71.
Series 3 (1965) projects # 83, 87, 93, 107, 110.
Series 4 (1965) projects # 153, 157.

BOARD OF EDUCATION, NEW YORK CITY. *Adventures in Biology.* Brooklyn, N.Y.: New York City Board of Education, 1962.

This is a collection of project ideas for high school biology students. Group 3, pages 117-154, includes project ideas in heredity and evolution.

DEMEREC, M. AND B. P. KAUFMANN. *Drosophila Guide.* Washington, D.C.: Carnegie Institution of Washington, 1965.

This is the most recent edition of a guidebook that has gone through 8 editions. It describes the techniques for culturing *Drosophila,* and suggests a series of experiments which the beginner can carry out. It is an excellent reference source, and a fine guide for project work.

TURTOX. *Service Leaflet # 15—The Culture of Drosophila Flies and their Use in Demonstrating Mendel's Law of Heredity.* Chicago, Illinois: General Biological Supply House, Undated.

One of a series of service leaflets describing various techniques. As its title implies, leaflet # 15 describes some of the fundamental techniques of *Drosophila* culture.

INDEX